Toward an Intercultural Natural History of Brazil

This volume presents the first extensive census of the surviving copies of the treatise *Historia Naturalis Brasiliae* in libraries worldwide and examines the book from a variety of interdisciplinary viewpoints.

The chapters in this volume are written by scholars from different fields of knowledge, including anthropology, botany, linguistics, literature, book history, medieval and early modern history, and art history. The chapters contextualize the treatise vis-à-vis its predecessors and contemporaneous works of natural history and examine its botanical, zoological, and linguistic accuracy and usefulness in the present day. Put together, the seven chapters of this volume present a kaleidoscope of possibilities of how to reinterpret Piso and Marcgraf's work within the dynamic context of knowledge-production about the "New" World in the early modern era, while also suggesting approaches to continue profiting from its subject matter in the present day.

Toward an Intercultural Natural History of Brazil offers essential reading on the *Historia Naturalis Brasiliae*, natural history, and Latin American history.

Mariana Françozo is an Associate Professor at the Faculty of Archaeology, Leiden University, The Netherlands. Her research stands at the intersection of anthropology and history and focuses on the collection and circulation of Indigenous objects and knowledge from Brazil to Europe, with special emphasis on the early modern period.

Routledge Studies in Global Latin America
Series Editors:

Peter Burke
University of Cambridge

Jorge Cañizares-Esguerra
The University of Texas at Austin

Linda Newson
School of Advanced Study, University of London

Mark Thurner
School of Advanced Study, University of London, and FLACSO

Routledge Studies in Global Latin America publishes critical, post-Area Studies scholarship that connects local histories with the global history of modernity. The editors are keen to publish in those areas where Latin or Iberian America has played a pioneering role in global history.

The First Wave of Decolonization
Edited by Mark Thurner

A History of Book Publishing in Contemporary Latin America
Gustavo Sorá

The Invention of Humboldt
On the Geopolitics of Knowledge
Edited by Mark Thurner and Jorge Cañizares-Esguerra

Toward an Intercultural Natural History of Brazil
The *Historia Naturalis Brasiliae* Reconsidered
Edited by Mariana Françozo

For more information about this series, please visit: https://www.routledge.com/Routledge-Studies-in-Global-Latin-America/book-series/RSGLA

Toward an Intercultural Natural History of Brazil

The *Historia Naturalis Brasiliae* Reconsidered

Edited by
Mariana Françozo

Routledge
Taylor & Francis Group

NEW YORK AND LONDON

First published 2023
by Routledge
605 Third Avenue, New York, NY 10158

and by Routledge
4 Park Square, Milton Park, Abingdon, Oxon, OX14 4RN

Routledge is an imprint of the Taylor & Francis Group, an informa business

Library of Congress Cataloging-in-Publication Data
Names: Françozo, Mariana de Campos, 1979- editor.
Title: Toward an intercultural natural history of Brazil : the Historia Naturalis Brasiliae reconsidered / edited by Mariana Françozo.
Other titles: Historia Naturalis Brasiliae reconsidered
Description: New York, NY : Routledge, 2023. | Series: Routledge studies in global Latin America | Includes bibliographical references and index.
Identifiers: LCCN 2022052717 (print) | LCCN 2022052718 (ebook) |
Subjects: LCSH: Marggraf, Georg, 1610-1644. Historia naturalis Brasiliae. | Piso, Willem, 1611-1678. Historia naturalis Brasiliae. | Marggraf, Georg, 1610-1644. Historia naturalis Brasiliae--Bibliography. | Piso, Willem, 1611-1678. Historia naturalis Brasiliae--Bibliography. | Natural history--Brazil.
Classification: LCC QH117.P673 T69 2023 (print) |
LCC QH117.P673 (ebook) | DDC 508.81--dc23/eng/20221221
LC record available at https://lccn.loc.gov/2022052717
LC ebook record available at https://lccn.loc.gov/2022052718

ISBN: 978-1-032-42472-9 (hbk)
ISBN: 978-1-032-42474-3 (pbk)
ISBN: 978-1-003-36292-0 (ebk)

DOI: 10.4324/9781003362920

Typeset in Sabon
by KnowledgeWorks Global Ltd.

Contents

Figures

Tables

Contributors

Alex Alsemgeest is Curator of Library Collections at the Rijksmuseum, Amsterdam. He has published articles on natural history, the history of libraries and collections, and the early modern international book trade with a focus on Dutch-Swedish relations. He is co-author, with Charles Fransen, of *In Krabbengang door Kreeftenboeken: De Bibliotheca Carcinologica L.B. Holthuis* (Leiden, 2016), which was shortlisted for the Jan Wolker Prize in 2016.

Prof. Dr. Tinde van Andel (PhD, Utrecht University, 2000) is Professor of Ethnobotany at Wageningen University and Professor of the History of Botany and Gardens, Clusius chair, at Leiden University. She is also a researcher at the Naturalis Biodiversity Center. Her research focuses on traditional plant use, historic herbaria, and botanic illustrations, with special emphasis on West Africa and the Caribbean. She has published numerous articles in journals such as *Science; Nature Plants; Social History of Medicine; Journal of Ethnopharmacology; Biodiversity and Conservation;* and *Economic Botany* among many others.

Jeroen Bos (MA, Institute for History, Leiden University, 2003) is an independent scholar who works on early modern Dutch (colonial) history. He is currently lecturer at the Department of History at the University of Groningen and Open Access Officer at Radboud University (Nijmegen). He has published articles and given public outreach lectures on various topics related to the Dutch overseas expansion, such as cartography, colonial botany, and book history.

Dr. Aline da Cruz (PhD, Free University of Amsterdam, 2011) is Associate Professor of Linguistics at the Universidade Federal de Goiás, Brazil. She specializes in Brazilian Indigenous languages, with special attention to language contact and the origin and expansion of Nheengatu (*língua geral*). She is the author of *Fonologia e Gramática do Nheengatú* (LOT, 2011).

Dr. Mariana Françozo (PhD, Unicamp, Brazil, 2009) is Associate Professor of Museum Studies, Faculty of Archaeology, Leiden University, the Netherlands. Her research stands at the intersection of anthropology and history and focuses on the collection and circulation of Indigenous objects and knowledge from South America and the Caribbean to Europe. She is the author of *De Olinda a Holanda: O Gabinete de Curiosidades de Nassau* (Ed. Unicamp, 2014). She is currently P.I. of the ERC Starting Grant Project *BRASILIAE: Indigenous Knowledge in the Making of Science.*

Dr. Walkíria Neiva Praça (PhD, Universidade de Brasília, 2007) is Associate Professor of Linguistics at the Universidade de Brasília, Brazil. She specializes in Brazilian Indigenous languages, with special attention to the language family Tupi-Guarani. She has published articles and book chapters on morphology and syntax of Tapirapé and Tupinambá; and on Indigenous education in Brazil.

Mireia Alcantara Rodriguez (MSc, Faculty of Archaeology, Utrecht University, 2015) is an ethnobotanist and currently a PhD candidate at Leiden University in the ERC Project *BRASILIAE. Indigenous Knowledge in the Making of Science: Historia Naturalis Brasiliae (1648).* With an ethnobotanical approach, she investigates methods of knowledge production in the early modern period. To this end, she compares plant uses documented in the HNB with modern day uses in Brazil. Additionally, she identifies the flora in the iconography created in Dutch Brazil to study the methods of plant collection by naturalists and artists in the colony, as well as the origins and current conservation status of the represented species.

Dr. Anjana Singh (PhD, Leiden University, 2007) is Assistant Professor of Early Modern South Asia and Global History at the University of Groningen, the Netherlands. Her research focuses on the Dutch East India Company (VOC) and on the social history of knowledge. She has contributed to the history of colonialism in South Asia and Indian Ocean studies. She is the author of *Fort Cochin in Kerala, 1750-1830: The Social Condition of a Dutch Community in an Indian Milieu* (Brill, 2010) and "Connected by Emotions and Experiences: Monarchs, Merchants, Mercenaries, and Migrants in the Early Modern World" in Felipe Fernandez-Armesto ed., *The Oxford Illustrated History of the World* (OUP, 2019).

Prof. Dr. Paul J. Smith (PhD, Leiden University, 1985) is Emeritus Professor of French Literature at Leiden University. He has published numerous books, articles, and book chapters on French literature, its reception in the Netherlands, French and Dutch fable and emblem books, literary rhetoric, and early modern zoology. On the subject of early modern natural history, he co-edited, with Karl Enenkel, *Early Modern Zoology: The Construction of Animals in Science, Literature and the Visual Arts* (2007), *Zoology in Early Modern Culture* (2014), *Emblems and the Natural World* (2017), and with Raphaële Garrod, *Natural History in Early Modern France* (2018).

Prof. Dr. Dr. Timothy D. Walker (PhD, Boston University, 2001) is Professor of History at the University of Massachusetts Dartmouth. He is the author of *Doctors, Folk Medicine and the Inquisition: The Repression of Magical Healing in Portugal during the Enlightenment* (Brill, 2005) as well as numerous articles and book chapters on the history of medicine in the Portuguese Empire; he is co-editor, with Harold Cook, of a special issue of the journal *Social History of Medicine* on "Mobilising Medicine: Trade & Healing in the Early Modern Atlantic World" (2013).

Dr. Annemarieke Willemsen (PhD, Radboud University Nijmegen, 1998), archaeologist and art historian, is Curator of Medieval Collections at the National Museum of Antiquities, Leiden. She has published numerous books and articles on Roman, medieval, and early modern material culture and everyday life, with a focus on children, play, education, and fashion.

Acknowledgments

This book was first imagined and discussed during a seminar organized at Leiden University in October 2016 and funded by a Leiden Global Interactions Profile Area Advanced Seminar Grant, for which I am thankful. While not all of the workshop participants eventually contributed to this volume, I wish to thank all of them for their participation in the event and for their enthusiastic support to this book project: Manuela Carneiro da Cunha, Tinde van Andel, Mireia Alcántara Rodríguez, Paul Smith, Amy Buono, Aline da Cruz, Eithne Carlin, Adrian Gomes, Anjana Singh, Timothy Walker, Jeroen Bos, Alex Alsemgeest, Neil Safier, and Michiel van Groesen. In the fall of 2017, I was Almeida Family-John M. Monteiro Memorial Fellow at the John Carter Brown Library, where I had the privilege of exploring my initial ideas about the *Historia Naturalis Brasiliae* (HNB) with the other fellows in residence, and particularly with Sara Guengerich and Barbara Mundy. In Providence, I enjoyed the support and friendship of Neil Safier and Iris Montero Sobrevilla, whom I thank for our ongoing intellectual exchanges about the HNB and so much more. I would also like to thank the LAGLOBAL network, particularly Mark Thurner, Jorge Cañizares-Esguerra, Miruna Achim, José Pardo-Tomás, and Juan Pimentel for their interest in and suggestions to my work. During the process of producing this book, Angela Hess and Csilla Ariese provided invaluable help and much-appreciated support. Last but not least, I thank all of the authors in this volume for their contributions and their patience.

This research is part of the ERC project *BRASILIAE. Indigenous Knowledge in the Making of Science*, directed by Dr. Mariana Françozo at Leiden University and funded by the European Research Council Horizon 2020 Research and Innovation Programme (Agreement No. 715423).

Mariana Françozo
Amsterdam, winter 2021/22

Introduction

Mariana Françozo[1]

This chapter introduces the main topic and the seven chapters of the edited volume *Toward an Intercultural Natural History of Brazil: The Historia Naturalis Brasiliae Reconsidered*. First, it situates the publication of Piso and Marcgraf's treatise in the context of the Dutch Republic and its global economic ambitions in the mid-seventeenth century. It then briefly reviews its reception history and its impact on science in order to explain the need and importance of yet another scholarly publication about the *Historia Naturalis Brasiliae*. Each one of the seven chapters is thus briefly summarized. This introductory chapter argues that a rich, multi-layered, and novel interpretation of this book can emerge from the interdisciplinary set of research questions and methodologies used in this volume's reconsideration of the treatise.

In 1912, American ichthyologist Eugene Gudger published an article in *Popular Science Monthly* titled *George Marcgrave, the First Student of American Natural History*.[2] The article is a biography of the naturalist, whom Gudger argues was the "… man who first of all essayed to make known to the old world the real natural history of the new."[3] The article presents a short reconstruction of Marcgraf's training as a naturalist, a brief description of his activities in Brazil, a reflection on his ichthyologic contributions, and much praise for his manuscript *Historia Rerum Naturalium Brasiliae*, published together with physician Willem Piso's work on tropical medicine under the title *Historia Naturalis Brasiliae* (henceforth HNB) by Elzevier in Leiden and Amsterdam in 1648.[4] Gudger claimed that practically all of the plant and animal species described and figured in this treatise – about 600 and 400 of them, respectively – were "new to science."[5] It remains otherwise unexplained what the author precisely means by these terms, but one can assume with some degree of certainty that Gudger had modern, post-Linnean standards of zoological and botanical description in mind.

Despite Gudger's rather anachronistic take on Marcgraf's work, the HNB did indeed make an impact on "Old" World science. Yet, Marcgraf was not the first to write about "New" World nature nor was Gudger the first to write about him – so much so that, only two years later, he published a

DOI: 10.4324/9781003362920-1

postscript in *Science* lamenting the fact that he had missed the publication, in 1908, of Brazilian historian Alfredo de Carvalho's article on Marcgraf, and adding some data and bibliographical reflections to his earlier text.[6] Carvalho, in turn, acknowledged the place of sixteenth-century French authors Jean de Léry and André Thevet as the first to describe Brazilian fauna and flora, and of Francisco Hernandez, commissioned by Phillip II, to have done so for Mexico. However, much like Gudger, Carvalho also firmly defended the point that Marcgraf was – almost one century later – the scholar who first started the "modern" study of Brazilian nature.[7] Both articles are representative of the overly positive tone frequently found in the historiography of Dutch Brazil, placing the Dutch period as a formidable alternative to the Portuguese colonization, or, as aptly put by Joan-Pau Rubiés, "a philo-Dutch counter-myth centered on the figure of Johan Maurits" that reasserts Dutch exceptionalism and modernity in the early modern era.[8] In this counter-myth, the HNB features as one of the finest material outcomes.

This is not to say that the HNB is *not* a remarkable product of Johan Maurits of Nassau-Siegen's colonial endeavors and editorial sponsorship. An in-folio tome of about 400 pages, containing hundreds of animal and plant descriptions and their images, the HNB impressed the early modern reader with an abundance of well-organized information about Brazilian nature. However, its uniqueness does not rest on its supposed first place in an extemporaneous chronology of European scientific explorers of "New" World nature, but rather on a complex combination of internal and external factors, including its physical qualities, its lavish illustrations, the careful program of distribution carried out by its patron and its publisher, the dynamics of the Dutch printing market in the early modern period, and, of course, the sheer amount and variety of natural history descriptions presented therein. Additionally, in the HNB, the commercial and political interests of the Dutch in the Americas are clearly addressed, adding a layer of political pamphleteering to its many functions.

To minimize confusion, in this volume we have chosen to consistently refer to the HNB as a whole as a "treatise" or a "tome;" its original division into two separately titled and paginated works by Piso and Marcgraf respectively is maintained by using the terms "part I" (Piso) and "part II" (Marcgraf); finally, the subsequent separation into "*libri*" are either translated literally as "books" or referred to by the more familiar "chapters."

In 1658, Willem Piso published a treatise named *De Indiae utriusque Re Naturali et Medica Libri Quatuordecim*, which is often erroneously considered as a second edition of the HNB.[9] In fact, Piso's volume is a rearrangement of many sections of the HNB. *De Indiae utriusque ...* is divided into two parts. The first one contains six chapters on tropical medicine alongside chapters on the plants and animals of Brazil. The former were authored by Piso and correspond to his work as published in part I of the HNB; the latter are taken from Marcgraf's *Historia Rerum ...* (part II of the HNB), but not attributed to him thereby suggesting that these studies had also been carried out by Piso. The second part of *De Indiae utriusque ...* is likewise composed by Marcgraf's two chapters on topography, meteorology, and on the inhabitants of Brazil and Chile, corresponding to the HNB's part II, chapter ("book") eight. Finally, this volume also includes Jacobus Bontius' natural history of Java in the East Indies, *De Medicina Indorum*,[10] thereby uniting the Dutch West and East Indies in one treatise and giving the book a global range.

While a detailed comparison of these two versions of the HNB is outside the scope of this introduction, it is important to briefly address the political and economic context of the Dutch Republic (or the United Provinces of the Netherlands) in which these treatises were created. In the mid-seventeenth century, the Dutch East- and West India Companies, with the support of the central government (the *Staten Generaal*), were busy trying to expand and consolidate their colonies abroad and to strengthen their commercial activities all around the world. Yet, much changed in the 10 years that separated the publication of the HNB and of Piso's *De Indiae utriusque ...* By 1648, the United Provinces had just gained their independence from the Spanish Empire after the Eighty Years' War and the HNB contains clear signs of attempts to confront and compare the Dutch overseas possessions and their economic potential to those of the Habsburgs. For instance, in the entries about sugar and manioc, Johannes de Laet, editor of the HNB and, importantly, also one of the directors of the West India Company (WIC), added an entire new chapter to compare Marcgraf's reflections on sugar and manioc to the writings of Francisco Ximenes on the same plants in New Spain. Both crops were essential to the economy of Dutch Brazil. Sugar was a cash crop and the main reason why the WIC conquered Brazil; as for manioc, Johan Maurits had made it mandatory for sugar-planters to devote a percentage of their lands to growing manioc and, in doing so, tried to solve the problem of hunger in the colony. In highlighting the methods of and riches coming from sugar production, and in drawing attention to the governor-general's solution to the food crisis in Brazil, De Laet also showed prospective colonists and the *Staten Generaal* how the colony was a viable source of profit. The HNB was therefore a guidebook to life in the colony just as much as a comparison between Dutch and Spanish overseas possessions; one that included a strong statement about the viability of Brazil for the Dutch.

The loss of Dutch Brazil in 1654 put a definite end to the public debate on Brazil and the Atlantic, which had peaked in the 1640s,[11] and by the time Piso's volume came out there was no reason to invest in propaganda for the WIC nor to praise the riches of Brazil. Therefore, in republishing (parts of) the HNB, Piso chose a different title (and different ensuing title page): one that reinforced Dutch commercial possibilities all over the globe, but particularly in the East Indies where the profits of international trade were to be found. In this sense, both treatises, while being extremely similar in textual terms, represent very different political projects of the Dutch Republic.

Despite Piso's effort to replicate the success of the HNB, it was the original edition that made history, partially due to the patronage of Johan Maurits and the strategic editorship and distribution by De Laet and the Elzevier publishing house. Similarly, as Neil Safier has pointed out, "part of this wide appeal had to do quite simply with the fact that the HNB was the sole natural historical text printed in the seventeenth century that focused primarily, if not exclusively, on South America."[12] Immediately after its publication, it became a much-cited, authoritative source on the fauna and flora of South America, so much so that the Swedish naturalist Carl Linnaeus used its names and descriptions of species to create part of his taxonomic system and received lavish praise in Diderot and d'Alembert's *Encyclopédie*.[13] Throughout the eighteenth century, scholars continued to refer to Piso and Marcgraf's work as the leading authority on the natural history of Brazil, providing material for contrast and comparison to nature elsewhere in the "New" World.[14] In fact, its impact – and physical presence – went beyond Europe and the Americas: a number of copies of the HNB were taken by James Cook and his crew for consultation during their voyages on board the *Endeavour*,[15] and its physical presence in historical libraries around the world proves the extent of its dissemination.[16] It is generally agreed that, at least in terms of the breadth and scope of identification and description of botanical species, the treatise was only superseded two centuries later by the *Flora Brasiliensis* of Johann Baptiste von Spix and Carl von Martius.[17]

Up until the present day, scholars continue to discuss the HNB from diverse disciplinary angles. From the perspective of life sciences, scholars have been using the treatise as a source of information on the biodiversity of Brazil, with recent scholarly literature using it for taxonomic identification and comparative studies between the presence, the naming, and the use of plants in colonial era and present-day Brazil.[18] Similarly, the HNB has been studied from a historical perspective as an object of inquiry in itself. In this sense, many are the publications praising its contributions to "New" World botany and zoology,[19] as well as those critically discussing the treatise from the perspectives of the history of science and medicine, and art history.[20] Marcgraf and Piso have each received due attention in biographical works.[21] Furthermore, there have been new studies on the astronomical section of the HNB, as well as on the expeditions and cartographical work of Georg Marcgraf.[22]

One may wonder, what else is there to be said about such a well-known tome? The aim of this edited volume is to reconsider the HNB as a multi-layered, complex compilation of experiences and knowledge, one that is better understood if looked at from multiple, sometimes contradictory, interdisciplinary viewpoints. Instead of considering the HNB as a masterpiece of Western science, the chapters in this volume interpret it as the product of global intercultural encounters in the early modern era. As such, chapters attempt to contextualize the treatise vis-à-vis its predecessors and contemporaneous works of natural history and to test its botanical, zoological, and linguistic accuracy and usefulness in the present day. Moreover, this volume hopes to suggest a model for scholarship on the history and historiography of knowledge, arguing that it is possible to write global histories of knowledge-production by concentrating on one individual, physically concrete object of study, which in turn can be deconstructed into a set of entangled parts or multiplied into various interdisciplinary questions. In this sense, and in order to reframe the HNB as a point of reference rather than a starting point, this edited volume contains seven chapters by scholars representing different fields of knowledge, including anthropology, botany, linguistics, book history, early modern and medieval history, and art history. Therefore, the chapters reflect the types of scholarly questions and methodologies typical of their authors' respective disciplinary trainings.

The volume starts with a chapter that introduces the making of the HNB from a comparative perspective. Singh and Françozo explore how both the HNB, on the natural history of Brazil, and the *Hortus Malabaricus*, on Indian botany, are products of Dutch colonial engagements with Indigenous knowledge-systems. By doing so, the chapter proposes that the HNB can be better understood when read alongside other such treatises of natural history produced in the seventeenth century Netherlands, following a specific pattern of information acquisition and later publication.

Chapter 2 moves from the Dutch to the Portuguese Empire and likewise reads the HNB comparatively, this time alongside Portuguese medical texts of the seventeenth- and eighteenth-centuries. Walker researches how each of them codified Indigenous knowledge and, although the author defends the superiority of the HNB vis-à-vis its Iberian counterparts, he also points out and explores the essential role of works such as Da Orta's in providing Piso and Marcgraf with a framework and a basis from which they could prepare for and imagine what they would find in the "New" World.

Chapter 3, by Alsemgeest and Bos, presents the first complete census of the existing copies of the HNB in public libraries worldwide, locating for the first time more than 300 copies of the treatise, including the identification of another 8 colored copies in addition to the 6 that were previously known. Based on data from the census, the chapter analyzes the patterns and particularities of the trajectory of this tome, making important points about its initial distribution and the distinction it later received. The chapter is accompanied by the census itself, published as an appendix towards the end of this volume.

In Chapter 4, Van Andel, Françozo, and Alcántara Rodríguez delve into one particular, important plant product of the tropics – the copaiba balsam – and review a series of early reports on this species and its medicinal properties. By exploring accounts from the sixteenth- and seventeenth-centuries, as well as contemporaneous pharmacopeia, the authors show the confusion that arose in the taxonomy, uses, and names for this plant, thereby highlighting how the careful study of the original historical documents associated with the HNB, such as the images in the *Libri Picturati* in Kraków, can greatly assist in the identification of plants and thus solidifying the usefulness of such colonial natural histories for present-day botanists.

Chapter 5, by Willemsen, refers to the medieval tradition of bestiaries to explore changes and continuities in the format and content of the HNB vis-à-vis the description of animals. The chapter defends the point that the treatise is as much rooted in tradition as it is modern, and, furthermore, adds evidence to the central role played by De Laet in the composition of the tome.

In Chapter 6, Smith offers an autoptic reading of the HNB focusing on the chapters about fish and birds. The chapter explores how, in his descriptions, Marcgraf sometimes closely follows, while at other times completely diverges from, the sixteenth-century naturalist tradition, thereby placing Marcgraf in conversation with Aldrovandi, Gessner, Belon, and Rondelet – and later on also comparing him to Willughby and Ray.

Finally, in Chapter 7, Cruz and Praça explore the Indigenous terminology used in the HNB as an entry point by which to expand an ongoing project of language documentation among speakers of languages of the Tupi-stock in contemporary Brazil. The chapter revisits the history of the use of Tupi as *língua-geral* in Portuguese America. It then presents the results of research carried out with three distinct Indigenous peoples in Brazil: the Apyãwa, Baré, and Tapeba. In doing so, the chapter proposes a present-day use for this centuries-old tome, aligning one of the material legacies of Dutch Brazil with contemporary efforts for language revitalization.

The main questions that cut across all contributions address the tensions between the treatise's modernity *versus* tradition; its practical usefulness as a guidebook *versus* its character as a diplomatic gift or a collectible tome; the co-existence of Marcgraf and Piso's eyewitness observations *versus* the recurring references to classic and earlier naturalist accounts, among others. Put together, the seven chapters of this volume present a kaleidoscope of possibilities of how to interpret the HNB within the dynamic context of knowledge production about the "New" World in the early modern era, while also suggesting approaches to continue profiting from its subject matter in the present day.

Notes

1 The research for this chapter is part of the ERC project *BRASILIAE. Indigenous Knowledge in the Making of Science*, directed by Dr. Mariana Françozo at Leiden University and funded by the European Research Council Horizon 2020 Research and Innovation Programme (Agreement No. 715423).
2 Eugene W. Gudger, "George Marcgrave, the First Student of American Natural History," *Popular Science Monthly* 81 (1912): 250–273.
3 Gudger, "George Marcgrave," 251.
4 Willem Piso and Georg Marcgraf, *Historia Naturalis Brasiliae: In qua non tantum Plantae et Animalia, sed et Indigenarum Morbi, Ingenia et Mores Describuntur et Iconibus supra Quingentas Illustrantur* (Leiden and Amsterdam: Elzevier, 1648).

The only translation of this treatise was published in the mid-twentieth century in Brazil in the form of two books that separate the HNB into its two parts: George Marcgrave, *História Natural do Brasil*, trans. José Procópio de Magalhães (São Paulo: Imprensa Oficial do Estado: 1942 [1648]); and Guilherme Piso, *História Natural do Brasil Ilustrada* (São Paulo: Companhia Editora Nacional, 1948 [1648]).

5 Gudger, "George Marcgrave," 261.

6 Eugene W. Gudger, "George Marcgrave, a postscript," *Science* 40, no. 1032 (1914): 507–509.

7 Alfredo de Carvalho, "Um Naturalista do Século XVII: George Markgraf 1610–1644," *Revista do Instituto Histórico Arqueológico e Geográfico Pernambucano* XIII (1908): 215.

8 Joan-Pau Rubiés, "Epilogue: Mythologies of Dutch Brazil," in *The Legacy of Dutch Brazil*, ed. Michiel van Groesen (New York, NY and Cambridge, UK: Cambridge University Press, 2014), 264–318.

9 Willem Piso, *De Indiæ utriusque Re Naturali et Medica Libri Quatuordecim* (Amsterdam: Elzevier, 1658). A Brazilian translation of the book came out as Guilherme Piso, *História Natural e Médica da Índia Ocidental* (Rio de Janeiro: Instituto Nacional do Livro, 1957).

10 Jacob Bontius, *De Medicina Indorum* (Leiden: Franciscum Hackium, 1642).

11 Michiel van Groesen, *Amsterdam's Atlantic: Print Culture and the Making of Dutch Brazil* (Philadelphia, PA: University of Pennsylvania Press, 2017).

12 Neil Safier, "Beyond Brazilian Nature: The Editorial Itineraries of Marcgraf and Piso's *Historia Naturalis Brasiliae*," in *The Legacy of Dutch Brazil*, ed. Michiel van Groesen (New York, NY and Cambridge, UK: Cambridge University Press, 2014), 169.

13 Peter J.P. Whitehead and Marinus Boeseman, *A Portrait of Dutch 17th Century Brazil: Animals, Plants and People by the Artists of Johan Maurits of Nassau* (Amsterdam: North Holland Publishing Company, 1989); Safier, "Beyond Brazilian Nature," 178.

14 See Willemsen in this volume.

15 Denis J. Carr, "The Books that Sailed with the Endeavour," *Endeavour* 7, no. 4 (1983): 194–201.

16 See Alsemgeest and Bos in this volume.

17 Carl von Martius, *Flora Brasiliensis* (Munich and Leipzig: R. Oldenbourg, 1840–1906). For a discussion of what "superseded" actually encompassed, see Safier, "Beyond Brazilian Nature."

18 Guilherme Garbino, Carla Aquino and Raone Beltrão-Mendes, "Marcgrave's Red-tailed Monkey: The Earliest European Depiction of a Titi Monkey," *Archives of Natural History* 48 (2021): 131–138, doi: 10.3366/anh.2021.0692; Mireia Alcantara-Rodriguez, Mariana Françozo and Tinde van Andel, "Plant Knowledge in the *Historia Naturalis Brasiliae* (1648): Retentions of Seventeenth-Century Plant Use in Brazil," *Economic Botany* 73, no. 3 (2019): 390–404, doi: 10.1007/s12231-019-09469-w; Mireia Alcántara Rodríguez, Isabela Pombo Geertsma, Mariana Françozo, and Tinde van Andel, "Marcgrave and Piso's Plants for Sale: The Presence of Plant Species and Names from the Historia Naturalis Brasiliae (1648) in Contemporary Brazilian Markets," *Journal of Ethnopharmacology* 259 (2020): 112911, doi: 10.1016/j.jep.2020.112911.

19 Maria F.T. Medeiros and Ulysses P. Albuquerque, "Food Flora in 17th Century Northeast Region of Brazil in *Historia Naturalis Brasiliae*," *Journal of Ethnobiology and Ethnomedicine* 10 (2014): 50, doi: 10.1186/1746-4269-10-50; Peter Wagner, "Das Markgraf-Herbarium," in *Sein Feld War die Welt: Johann Moritz von Nassau-Siegen (1604–1679)*, ed. Gerhard Brunn and Cornelius Neutsch (Münster: Waxmann, 2008), 233–245; Marinus Boeseman, "A Hidden Early

Source of Information on North-Eastern Brazilian Zoology," *Zoologische Mededelingen Leiden* 68, no. 12 (1994): 113–125; Peter Whitehead, "Georg Marcgraf and Brazilian Zoology," in *Johan Maurits van Nassau-Siegen 1604–1679*, ed. Ernst van den Boogaart (The Hague: The Johan Maurits van Nassau Stichting, 1979), 425–471; Peter Whitehead, "The Original Drawings for the *Historia Naturalis Brasiliae* of Piso and Marcgrave (1648)," *Journal of the Society for the Bibliography of Natural History* 7, no. 4 (1976): 409–422, doi: 10.3366/jsbnh.1976.7.4.409.

20   Safier, "Beyond Brazilian Nature"; Harold J. Cook, *Matters of Exchange: Commerce, Medicine, and Science in the Dutch Golden Age* (New Haven, CT: Yale University Press, 2007); Timothy D. Walker, "The Medicines Trade in the Portuguese Atlantic World: Acquisition and Dissemination of Healing Knowledge from Brazil (c. 1580–1800)," *Social History of Medicine* 26, no. 3 (2013): 403–431, doi: 10.1093/shm/hkt010; Junia Furtado, "Tropical Empiricism: Making Medical Knowledge in Colonial Brazil," in *Science and Empire in the Atlantic World*, ed. James Delbourgo and Nicholas Dew (New York, NY: Routledge, 2007), 127–152; David Freedberg, "Science, Commerce, and Art: Neglected Topics at the Junction of History and Art History," in *Art in History. History in Art*, ed. David Freedberg and Jan de Vries (Santa Monica: The Getty Center Publication Programs, 1991), 376–428; Julie Hochstrasser, "Human Nature: Observing Dutch Brazil," in *Engaging with Nature: Essays on the Natural World in Medieval and Early Modern Europe*, ed. Barbara Hanawalt and Lisa Kiser (Notre Dame: University of Notre Dame Press, 2008), 155–199.

21   Elke Pies, *Willem Piso (1611–1678)* (Düsseldorf: Interma-orb Verlagsgruppe, 1981); Rebecca Parker Brienen, "Georg Marcgraf (1610–c.1644): A German Cartographer, Astronomer and Naturalist-Illustrator in Colonial Dutch Brazil," *Itinerario* 25, no. 1 (2001): 85–122, doi: 10.1017/S0165115300005581.

22   Oscar Matsuura and Huub Zuidervaart, "America's Earliest (European-style) Astronomical Observatory," in *Scientific Instruments in the History of Science: Studies in Transfer, Use and Preservation*, ed. Marcus Granato and Marta C. Lourenço (Rio de Janeiro: Museu de Astronomia e Ciências Afins, 2014), 33–52; Ernst van den Boogaart and Rebecca Parker Brienen, eds., *Informações do Ceará de George Marcgraf (Junho-Agosto de 1639)* (Rio de Janeiro: Index, 2002).

1 Locating Knowledge in Early Modern Brazil and India

A Comparative Study of *Historia Naturalis Brasiliae* (1648) and *Hortus Malabaricus* (1678–1693)[1]

Anjana Singh and Mariana Françozo

Introduction

The emergence of a print culture in early modern Europe, focused on the study of nature, was deeply linked to information gathering and knowledge production in numerous parts of the world. Historicizing the discipline of knowledge production and history of science from the perspective of numerous cultures is indispensable. Taking up two examples of tomes that were published in the seventeenth-century Low Countries, in this chapter we aim to compare and connect the processes of early modern knowledge production from a post-colonial perspective. The *Historia Naturalis Brasiliae*, published in 1648 (henceforth HNB) and based on colonial Latin America (Brazil), and the *Hortus Malabaricus*, published between 1678 and 1693 (henceforth HM) and based on South Asia (Malabar, India), serve our purpose of studying historical processes of knowing the natural world, gathering information about it in an intercultural context, and subsequently publishing natural history treatises.

While several scholars continue to embrace the idea of "the rise of Europe," others called for a review of this subject that led to a new set of history writing from revised non-Eurocentric perspectives.[2] With regard to the production of knowledge in the early modern period, it has been argued for the need of an expanded view on the making of modern science in intercultural contexts.[3] The globalization of knowledge in the "Old" and "New" World led to classifying, controlling, and selling it as "European" since early modern times. This has been a much studied and debated phenomenon.[4] Most recently in the history of the global interaction and construction of knowledge, especially that of history of science, there has been a call to think about history of knowledge without the idea of the "Scientific Revolution."[5] The historiography on knowledge in the Dutch expansion in the Indian Ocean and the Atlantic world is also growing. This set of historiography brings about knowledge production that depended on the Dutch trading networks of the seventeenth and eighteenth centuries. These works explore how the circulation of people

DOI: 10.4324/9781003362920-2

and products led to the production of knowledge in Eurasian and Latin American societies, focusing on epistemic changes in historiography, geography, religion, and philosophy.[6]

The aim of this chapter is to analyze the process by which Dutch men in service of the West India Company (WIC) and Dutch East India Company (VOC) gathered local knowledge in Brazil and South Asia and brought them into the context of knowledge production in the Low Countries. How and why did such remarkable treatises come into being? We shall analyze the different stages of information gathering, their taxonomy, illustration, and description, leading to the final production of the tome. We will assess Dutch overseas expansion and the impact of trade and colonial empire-building in global knowledge production. How did knowledge production on nature within the Dutch network of traders, scholars, soldiers, artists, and ordinary men and women come about? Studies in the history of medicine in particular have emphasized the contribution of Indigenous and local peoples in providing information and sets of practices that were incorporated into a Western body of knowledge, particularly so in the case of botany.[7] These studies have not, however, taken such knowledge as comparable to European science. Indigenous peoples remain relegated to the place of intermediaries, cultural brokers, or informants. In this chapter, we will attempt to focus on knowledge present in the HNB and the HM as a way of locating non-European epistemologies and worldviews. We lean on Chris Bayly's definitions on information and knowledge: "information implies observations perceived at a relatively low level of conceptual definition, on the validity of whose claims to truth, people from different regions, cultures, and linguistic groups might broadly agree. Knowledge implies socially organized and taxonomized information, about which agreement would be less sure."[8]

With the arrivals of Christopher Columbus, Pedro Álvares Cabral, and Vasco da Gama on the coasts of America and South Asia in the late fifteenth century, the scene was set for a global and comparatively accelerated circulation of biological material and information.[9] This was particularly beneficial for the production of knowledge in Europe as more commodities flowed from Asia and the Americas to Europe, rather than the other way around. Flow of goods and commodities from Europe to other parts of the world started only after the industrialization process and is a feature of what is referred to as the modern world. While several historians have described the information gathering and bringing of botanical knowledge of "Old" and "New" World into the context of knowledge production in Europe, it remains to be researched how exactly the process took place in an intercultural global setting. In the next two sections, we present two examples of early modern treatises about the natural worlds of South Asia and Brazil, which were created within the context of Dutch expansion. In the two sections following this, we have outlined in detail the process of knowledge creation and attempted to locate local knowledge in them. In the concluding section, we argue that multiple knowledge systems converged in print form in Amsterdam. This global connected perspective enables us to demonstrate the polycentric nature of the emergence of knowledge on nature in the early modern world.

The *Historia Naturalis Brasiliae* (1648)

In 1648, Johannes de Laet published Willem Piso and Georg Marcgraf's HNB, a treatise in two parts, containing 12 "books," on the botany and zoology of Brazil. In an encyclopedic format, it brings together information about the natural world, linguistics, astronomy, and geography of South America as understood and experienced by multiple groups of people, including Luso-Brazilians, Tupi and other Indigenous Brazilian peoples, Africans, and Afro-Brazilians. Its method of construction embodies the intercultural connections that shaped practices of knowledge production in colonial settings across the globe, and is one of the most important published examples of such from Brazil.

This beautiful tome is part of the legacy of the historical period now commonly known as "Dutch Brazil": the quarter of a century when the Dutch WIC managed to conquer and control a significant portion of the sugar-producing northeastern captaincies of Portuguese America (1630–1654). During those years, the WIC was able to take over the production chain of sugar from cultivation and processing to commerce and distribution in Europe. The Dutch colony was governed at first by a body of colonial administrators hired by the WIC, but in 1637 it appointed Johan Maurits of Nassau-Siegen (1604–1679) to be governor-general of the colony. He was a German count, descendant of the leader of the Dutch Revolt against the Spanish Habsburgs William the Silent (1533–1584), with successful military experience in the army of the Dutch Republic. He was, likewise, remarkably interested in the arts and sciences, so much so that he personally hired a group of artists and scientists to accompany him to the "New" World. The task of Johan Maurits' entourage, which included aforementioned naturalists Willem Piso and Georg Marcgraf, as well as painters Frans Post (1612–1680) and Albert Eckhout (ca.1610–1666) among others,[10] was to observe, collect, study, and register Brazilian fauna, flora, and inhabitants.

The excursions into the hinterland to collect information included members of the Dutch colony and employees of the WIC, often accompanied by Indigenous Brazilians. At least the first one of those expeditions did not have exploratory or knowledge-gathering aims: as Ernst van den Boogaart and Rebecca Parker Brienen have convincingly shown, the goal of the voyage to Ceará in 1639 was to capture and enslave Indigenous people to work for the Dutch colony. The only still extant piece of Marcgraf's diary – identified and translated by Van den Boogaart and Brienen – contains a brief description of this 1639 expedition. Marcgraf writes short sentences describing the main activities of the day, the distances traveled, and often small notes on the animals they encountered on the way. He mentions pigs, an ostrich, a few opossums, and a jaguar.[11] The interaction with Indigenous peoples is also clearly mentioned in this diary excerpt, as Marcgraf explains how the Brazilians assisted in all types of tasks during the expedition: on a certain occasion, the Brazilians tried to capture an ostrich but were unable to hold the animal, which shoved them to the ground and ran away.[12] It is very likely that, even on an expedition to capture and enslave people, Marcgraf made notes that were later included in the HNB. However, as Safier points out, "the broader context for these excursions [...] is absent from the text itself as it describes the natural characteristics of a New World."[13]

A close look at the HNB reveals the extent to which its authors, and its editor Johannes de Laet, relied on or discussed pieces of Indigenous and local knowledges in its various chapters. In Part I of the treatise, Willem Piso's treaty on tropical medicine, the practical usefulness of Indigenous knowledge in treating diseases is essential. The chapter on poisons and antidotes is a compelling example thereof. Piso tried to obtain information from local populations but apparently only succeeded in getting accounts from Indigenous collaborators. Historian Junia Ferreira Furtado states that "Piso does not denigrate native knowledge about American nature because he realizes that they could be useful to European medicine."[14] Moreover, she argues that Piso separated the practical knowledge he learned from Indigenous peoples from what he thought were their "beliefs and superstitions," thereby suggesting a division between practical or empirical pieces of information and full knowledge systems. Likewise, historian H. Carneiro claims that Piso's prejudice kept him from learning extensively and properly from his Indigenous informants: for instance, due to his discrimination, the Dutch doctor would have failed to understand the practical efficiency of painting one's body with the black ink of the *jenipapo*, which would protect the skin from insects and from the sun.[15] While Piso's prejudice against Indigenous cosmologies and ways of life may be apparent in his writing, the very fact that he had to depend on their medicinal practices to conduct his work in the tropics and, moreover, the fact that he reports on them in his writing, is strong evidence of the importance and continued presence of Indigenous knowledge in the creation of Western medicinal practices. This argument has been put forward by Timothy Walker in regards specifically to Portuguese medical accounts about South America: "Indigenous peoples of Brazil thus made important contributions to 'Western' medicine during the early modern period, but typically did so anonymously and indirectly through European intermediaries, who often failed to discuss the original human sources for this knowledge."[16] At specific points, the HNB hints at how these Indigenous and local practices were incorporated in everyday life by the Dutch in Brazil. For instance, the roots of the *caapeba* herb (*Piper marginatum*) were then considered excellent against kidney stones and Marcgraf reports that "a Portuguese man used to give it to Mr. Vander Dussen, with great results."[17]

Part II of the HNB, Marcgraf's eight books on the botany and zoology of Brazil, includes abundant examples of how Indigenous practices were translated and made comprehensible to the European readership of this treatise. Safier argues that Marcgraf's regular use of Indigenous terminology in the HNB contributed toward "establishing the reputation of the HNB as a reliable source."[18] More than just citing names, however, Marcgraf in fact compiled a catalogue of Indigenous uses of diverse plants and animals. In its edition by De Laet, the treatise was transformed into one of comparative botany and zoology, contrasting examples from Marcgraf's work in Brazil with that of earlier scholars in Spanish America.

Most examples of Marcgraf's engagement and research with Indigenous informants can be found in the botanical section of the HNB, namely, Marcgraf's first three chapters (on the herbs, the plants, and the trees of Brazil). In those chapters, one can read and learn about multiple practical uses of diverse plants and identify Marcgraf's amazement with some of the skills shown by the local populations. For instance, he describes how Indigenous peoples make fire without using stones: "… without hitting the stone with a steel instrument, [they] take a dry piece of stem or root of this tree [*ambaiba*]; make a small hole on the ground, introduce a pointy stick into it and move it around, holding that piece of stick firmly with the feet and applying dried leaves of trees or cotton into the hole. So [they] spark the fire as they please, and thus everywhere [they] can light the stove."[19] Marcgraf paid attention to Indigenous ways of life in many aspects. He reports on multiple uses of the same plants: not only is the trunk of the *jataboca* tree used to build the walls of houses, it is also carved out in the form of a large vessel to transport water to faraway and desert places.[20] The fruit of the calabash tree or *cuiete* (*Crescentia cujete*), the author notes, is frequently used as "plate, cup or bottle."[21] The ripe fruit is thrown into hot water and left to cook thoroughly. Then, they "perforate it to make a hole. If you want to cut it in the form of a plate or any other [form], you place a large cord around the cooked fruit, disposing it in the format that you want to get, then apply some hits to the cord with a wooden hammer and so the fruit is cut off." In a more or less confessing tone, Marcgraf goes on and comments that "if we try the same technique with a knife, it will be a useless waste of time."

Examples of Indigenous food practices are also to be found in the HNB, as well as some hints as to how they can be useful to Europeans. Marcgraf writes about the potato, the *ietuca* (Indigenous term) or *quiquoaquianputu* (in the Congo language), as a delicious edible item that the Indigenous peoples ferment with water into a drink.[22] The *quiya uca*, or pepper, is used as a spice in the preparation of meals "just like we use salt."[23] The Indigenous Brazilians mix it with fish and *farinha* (manioc flour) and take it on long trips. Out of the cashew fruit or *acaiaiba* (*Anacardium occidentale*), the Indigenous Brazilians make a type of wine; the fruit's nuts are used to count years of life, one cashew nut per year, and the tree's wood to make canoes.[24] Many other examples can be listed, including references to body paint with the juice extracted from certain fruits and the making of feathered ornaments with the feathers of *guaras* (*Scarlet ibis*) and parrots.[25]

The author and the editor also carefully indicate the plants and animals that were transported from other parts of the world to Brazil, and particularly from western Africa, mainly Angola and Congo, from which the Portuguese and later the Dutch continuously imported enslaved people to work in plantations. Marcgraf didactically explains, in the case of the *sesamum*, that "it does not have a special Indigenous name for it is not [native to] this region, having been brought from Africa by the Portuguese."[26] The HNB provides a good measure of the influx of West African species into the natural world of Brazil. The *quiyaqui* or chili pepper (*Capsicum frutescens*) came originally from Angola.[27] The *inaia guacuiba*, called originally *ejaquiambutu* in the Congo, is the infamous coconut tree, whose fruit is called *inajaguacu* by the Indigenous Brazilians.[28] The author goes on to explain its production and mentions that in 1640, about 300 men were employed in the transportation of such trees, already grown and 24 years of age, to the gardens of Mauritsstad. He furthermore mentions that the coconut shell is excellent to make drinking vessels. Marcgraf certainly saw and possibly drank from one of those coconut shells. There is evidence of a few coconut shells, beautifully carved out with native Brazilian themes, having belonged to the collection of Johan Maurits.[29] Marcgraf also alludes to a species of chicken brought from Sierra Leone to Brazil, as well as a number of species of monkeys from Guinea and the guinea pig.[30]

While Marcgraf pointed out the presence of West African animals and plants in Brazil, Johannes de Laet, editor of the treatise, did the work of adding dozens of notes comparing Marcgraf's findings with those of other, earlier naturalists who had been in the Americas. Especially the first three chapters of Marcgraf's part contain abundant notes referring to Francisco Ximenes' *Quatro Libros de la Naturaleza, y Virtudes de las Plantas, y Animales*.[31] In these notes, De Laet compares Brazilian Indigenous knowledge and use of plants to those of what he calls "Mexicans," or the inhabitants of "the islands," that is, the Caribbean, as described by Ximenes. These comparisons not only focus on terminology, but also take into account the Indigenous practical uses of elements. The aloe vera, or *caraguata* for the Indigenous Brazilians, is, according to De Laet, called *maguey* or *metl* by the Mexicans.[32] The plant *quiya*, which includes different species of pepper, is called *chili* by the Mexicans and *axi* on the Caribbean island of Hispaniola.[33] The use of plants and fruits to make paint is a noteworthy example of botanical and ethnographic comparisons. For instance, according to Marcgraf the Indigenous Brazilians use the *urucum* plant – in Latin, *Bixa Orellana* – to make both a drink against poisons and a red paint with which they adorn their bodies and the vessels made out of calabash.[34] De Laet adds a note reporting that, according to Ximenes, this plant is called *achiotl, changuaricam,* or *pomaquan* in New Spain and the Caribbean; the paint which is made from it called *roucu* by the Mexicans while the inhabitants of the (Caribbean) islands use it to paint their body in an elegant manner.[35] Likewise, De Laet compares Marcgraf's writings on the *jenipapo* tree with Ximenes' information on the *xahuali*. While Marcgraf makes no note of its use by Indigenous Brazilians, De Laet adds that "the barbarians, in their feasts or when they go to war, paint themselves with this liquid [made of the fruit of the tree] so that they would look more dangerous to their enemies."[36]

Food customs receive the same comparative addendum by the editor of the HNB. For instance, while Marcgraf describes in detail the process of growing, harvesting, and processing manioc by the Indigenous Brazilians and by the Portuguese in Brazil, De Laet reports that, in New Spain, the same root is called *quauhcamotli* and in the Caribbean, it is known as *yucca* and is used to make a bread called *cassava*.[37] Johannes de Laet inserted these comparative notes almost exclusively on the botanical sections of the HNB, with the exception of the description of the bird *urubu* (vulture), which according to him is called *tzopilotl* by the Mexicans and *aura* by "the others" (Caribbean Indigenous peoples).[38] In one of his few ethnographic observations, De Laet recounted that Spanish men in New Spain claimed to have been cured of venereal diseases by eating vulture meat. The ethnobotanical study that Marcgraf had compiled with information on the Brazilian natural world and the species brought from West Africa was thereby expanded by De Laet in a continuous comparison with examples from Spanish America.

The *Hortus Malabaricus* (1678–1693)

The Hortus Malabaricus is attributed to the efforts of Hendrik Adriaan van Reede, a Dutchman born in Utrecht in 1636. He joined the service of the VOC in 1657.[39] Not much is known about his early life and career in the VOC. In 1661 and 1662, Van Reede was among the VOC personnel who fought the Portuguese in Cochin, on the southwest coast of India.[40] The VOC had laid siege to the coast of Malabar and wanted to drive the Portuguese away in order to have access to trade in Cochin. It is currently known as Kochi and was an important and strategically located port city in the territories of the raja of Cochin.[41] The Portuguese had a settlement there since 1500.[42] It was valued by Europeans for its rich hinterland that produced pepper. The Dutch operated there between 1663 and 1795. It was then taken over by the British.[43]

Van Reede was made commander of Malabar in 1669. He held the office for seven years, i.e., till 1676. Van Reede was impressed by the region's rich flora from the very first moment that he saw it. Although, as expected, he first concentrated on the political and commercial matters of establishing the VOC there, he remained inquisitive of the flora and was interested in gathering more knowledge about it. In 1674, Matthew of St. Joseph, a missionary experienced in the making of illustrated manuscripts on medicinal plants, arrived in Cochin and the idea of compiling a book, the HM, came into being. Matthew had compiled the *Viridarium Orientale*, which he wanted to publish after a professor of Oriental Languages at Leiden had worked on it. This plan was never realized and the manuscript ended up in Italy, where Giacomo Zanoni, professor of botany at Bologna, published parts of it. Notes and drawings of *Viridarium Orientale*, compiled by Matthew of St. Joseph, served as a template for the first version of the manuscript of the HM. When the German naturalist Paul Hermann (1646–1695), who was collecting information on the flora of Ceylon, visited Cochin, a second and more definitive version of the HM came into being. Hermann served the VOC as a medical officer from 1672 to 1677. After his return to the Dutch Republic in 1679, he held a chair of botany at Leiden University, where he created an exceptionally fine botanical garden, the *Hortus Botanicus.*

Van Reede was a foreigner in Malabar. He had very little knowledge about the region, and most of his time and resources were spent in the service of the VOC. His main task would be having pepper and other commodities collected from the hinterland at agreeable prices from the local agents and corresponding with Ceylon, Batavia, and the Netherlands on the company's affairs on the Malabar. He had to gather information on rival Europeans, like the Portuguese, the French, the Danes, and the English, all of whom were actively competing in the region for political allies and commercial agents. A large part of his time would go into taking care of the affairs of the VOC, its personnel, forts, etc. Van Reede's interest in Malabar's plants was purely personal. Yet he also managed to use all possible resources of the company to gather information about regional plants, most of all through his position as commander, his office, ink and paper, and the personnel of the VOC. He invited locals to share their knowledge about plants with him and specially requested medics to visit him. He collected information on the medicinal qualities of the plants, their growing seasons, and other characteristics. Many of his informants were learned men and physician healers (*vaidyas*) of Malabar who had knowledge of plants and their healing powers. He appreciated the knowledge of the Brahmins with whom he came in contact; sent out messages that anyone who had any information about plants or access to plants could visit him; and wrote letters to princes and chiefs in an attempt to collect all oral and written botanical knowledge as well as specimens of the plants of Malabar and the Konkan Coast. Van Reede had become a trusted friend of Vira Kerala Varma, the Raja of Cochin, who must have supported his initiative. Van Reede himself traveled to Travancore and met people there to gather information from the hinterland.

One of Van Reede's informants was Itty Achuthan, a traditional vaidya at that time, who provided knowledge on medicinal plants. From the notes of Van Reede, we know that Itty Achuthan had hereditary palm leaf manuscripts. This is our only evidence of ethnomedical knowledge that circulated in Malabar at that time. We know that Itty Achuthan carried a palm leaf manuscript with him, but Van Reede does not give any further information on the name, content, author, or language of the text. Information about plants and their medicinal properties were written down and preserved in palm leaf *olas* and they were most probably handed down from one generation to the next; this knowledge-based society limited the access to knowledge to the ruling elite and priestly class, thereby limiting its circulation.

Another set of informants that Van Reede met and mentions in the HM were three Konkani priest physicians, Vinayaka Pandit, Ranga Bhatt, and Appu Bhatt. These three Brahmins had access to the *śāstrás*, the classical knowledge system of South Asia. Van Reede does not mention any books that the Brahmins read or brought with them to the meetings, but he emphasizes that they were exceptionally quick in delivering oral information on plants once they were told their names. Van Reede was impressed by their memory. The Brahmins thus had an orally transferred knowledge of plants and their characteristics or healing properties, which they had received either through their families or from their Brahmin teachers and was passed from one generation to the next. In this way, they supplemented the information given by Itty Achuthan.

In South Asia, efforts to locate manuscripts that could inform us about their systems of knowledge has yielded no results. Traditionally, knowledge in South Asia was mostly held and information passed on from one generation to another orally, and often within the family. Although it was not a highly literate society, it was acutely aware of literacy. Oral as well as written information moved swiftly through the medium of merchants, pilgrims, soldiers, or marriage parties. Knowledge was unevenly distributed within society; families and communities among the religious elites attempted to guard knowledge and reserve it for their descendants. Many influential groups recorded information in scripts and dialects which only a few people could understand.[44]

Thus, the exploration of nature, collecting useful and reliable information about it, and preserving and circulating knowledge about it in oral and written form was part of the Malabar intellectual milieu. Knowledge was openly shared with those interested in the field, even foreigners. No doubt, within different peoples of Malabar, caste formed opaque walls though which communication was limited. The people of Malabar had found different ways of knowing about nature. And they had found different ways of preserving that knowledge. It was a well-established long tradition of knowledge creation that had led to what Van Reede ultimately collected from them.

How Van Reede was able to gather information from different communities of Malabar who would normally not share space and information with each other due to caste and class restrictions is really remarkable. Van Reede brought people together and created for himself a network of informants that would contribute to the compiling of information from different systems of learning, preserving, and circulating knowledge. People who spoke Malayalam, Konkani, Portuguese, and Dutch and those with knowledge of Sanskrit, Arabic, and Latin were organized in a complex system of translation and transliteration so that all possible information on the names, characteristics, growing seasons, seeds, fruits, flowers, roots, and leaves, along with the medicinal values of the numerous plants and trees was written down. Van Reede hired draftsman to sketch precise images of the seeds, fruits, flowers, leaves, stems, and roots of the plants and trees. All plants were described and illustrated with their local Malayalam names, written in Roman, Malayalam, and Arabic scripts. In most cases, their Konkani, Portuguese, and Dutch names were also given. Van Reede thus had benefited from the knowledge and skills of many assistants and collaborators in Malabar. Some of them were Europeans, others mixed (*mestizos* and *castizos*), and many were local people of Malabar who contributed particular information to the making of Van Reede's manuscript. Perhaps Van Reede had learnt about a system of plant classification that was used by the Brahmins or by Itty Achuthan, who might have classified plants into different groups according to the local system(s) of classification. But this cannot be confirmed as we do not have Van Reede's personal notes and therefore we cannot report on Malabar's local plant classification system.[45]

In 1676, Van Reede was transferred to Batavia as an extraordinary member of the Council of the Indies. Thus, the first stage of the book-making process, i.e., the collection of data, came to an end. In Batavia he worked on the manuscript with the assistance of, among others, Willem ten Rhijne who was also an employee of the VOC. Ten Rhijne was a physician and had an interest in plants. He had worked in Deshima, Japan, where he had created a network of people who informed him about Japanese medicine and systems of healing. In 1676, he too was transferred to Batavia, where he met Van Reede. Van Reede and Ten Rhijne thus had a common interest: collecting knowledge in Asia about plants, their medicinal values, and other systems of healing. Ten Rhijne went on in 1683 to publish a treatise, which became the first detailed Western study on the art of using needles to cure bodily ailments. In 1677, Van Reede returned to the Dutch Republic. The information he had gathered in Asia had convinced him of its importance and he wanted to publish it in the Dutch Republic. The making of the HM now entered the next stage: preparation for publication in Europe.

Once back in the Dutch Republic, Van Reede took distance from his VOC network. He aspired to a career in politics and bought honorific titles for himself and his family name. Furthermore, he started to compile the information and drawings he had collected on the flora of Malabar as he aimed to publish his manuscript and drawings. Van Reede got in touch with Arnold Seyn, professor of botany at Leiden. As editor, Seyn reorganized the information according to the classification system prevalent in Europe at that time. Johannes van Someren and Jan van Dyke agreed to publish the books. Between 1678 and 1693, in total 12 volumes of the HM were published, with many challenges in between.

The first two volumes were published in 1678 and 1679. The first volume incorporated translated testaments of authenticity about knowledge from Malabar. The two volumes joined the vast corpus of knowledge about flora and fauna that had steadily grown in Europe, especially in the Low Countries since the printing press had been introduced and the study of nature popularized. The Low Countries and larger Europe were not only publishing on nature in Europe, but all over the globe. In 1681, Van Reede signed new publishing contracts and reorganized the editorial team. Johannes Munnicks, professor of botany at Utrecht, became editor, while Jan Commelin, an amateur botanist, continued as commentator. In 1682 the third volume was published. It had an extensive preface where Van Reede explained the aim, scope, and genesis of the work along with his impressions and feelings about the people and flora of Malabar. In 1684, Van Reede once more received a chance to travel to Asia for the VOC. He took up the post of commissioner general of Western Quarters. While Van Reede toured the Western Quarters, he also continued to gather botanical information especially at Cape Town and in Ceylon. Using the VOC's shipping, he sent plants and seeds from all these places to Amsterdam's Municipal Garden. In India, he traveled throughout the west and east coast of the peninsula up to Bengal. In December 1691, Van Reede died on board a VOC ship between Malabar and Surat, where he remains buried.[46] Nine more volumes of the HM were thus published posthumously. His original extensive notes and drawings that he had collected during his first stay in Malabar remained the basis. Different people co-edited the volumes, the titles mention: vol. I Joannes Caesarius and Arnoldus Syen, vol. II Joannes Caesarius and Joannes Commelinus, vols. III–V Joannes Munnicks and Joannes Commelinus, vol. VI Theodorus Janssonius van Almeloveen and Joannes Commelinus, vols. VI–XII Joannes Commelinus and Abraham van Poot. In 1693, the last of 12 Latin volumes appeared in print. All volumes state as author, acknowledge, or pay tribute to the work and efforts of Van Reede. In total, the 12 volumes contain the illustrations and descriptions of 742 plants of Malabar, belonging to 691 modern species.

Three editions of the HM from the early modern period are known. The first and original Latin edition of 12 volumes in folio was published in Amsterdam between 1678 and 1693. A reissue of this edition with new title pages was printed in Amsterdam in 1686. The second edition is a Dutch translation of the first two volumes published as *Malabaarse Kruidhof*, also printed in Amsterdam in 1689.[47] This was supposed to be a more popular version, which would sell numerous copies. A reissue of this popular and marketable Dutch edition with new title pages appeared in The Hague in 1720. The third edition came out in 1774. It was a modified Latin version, in quarto, of only the first volume, entitled *Horti Malabarici Pars Prima*, edited and annotated by John Mill and published in London. Van Resandt incorrectly states that it was also translated into English.[48] Since then, it has been as late as 2003 when K.S. Manilal published the first set of 12 volumes in English; in 2006, he published a complete set in Malayalam.[49] In this protracted way, with the efforts, vision, knowledge, and understanding of many different people, this set of 12 remarkable volumes of the HM came into being. It has a long history of circulation between Europe and Asia. No doubt, its impact has been global.

Production and Knowledge Creation in the Low Countries

A culture of collecting information on nature developed in the seventeenth-century Low Countries following the examples and practices of other European empires whose overseas possessions predated those of the Dutch. With maritime connections established with the "Old" and "New" Worlds, information flowed into the Low Countries at a far faster pace than ever before. Along with the print culture, in the early modern period, there emerged a culture of exhibiting exotic plants from around the world by setting up botanical gardens, such as those in Leiden and Amsterdam. The networks established by the VOC and the WIC contributed to this by shipping information as well as plants, seeds, fruits, animals, etc. to the Low Countries. Both companies pursued a policy of encouraging their personnel to gather information on tropical flora and fauna, especially medicinal and edible plants. These would be useful for the well-being of their servants, as well as profit making. The VOC and WIC were bio-prospecting agencies of the early modern period. Naturalists such as Georg Marcgraf and Willem Piso served the WIC and explored Latin America, while Georg Rumphius (1627–1702) and Paul Hermann explored the islands of Ambon and Ceylon for the VOC. Van Reede's works were completely in line with the policy of the company concerning the provisions of medicaments and the search for useful plants. The same can be said for Marcgraf's and Piso's work that resulted in the HNB. Rumphius' *Het Amboinsche Kruidboek* or *Herbarium Amboinense*, a catalogue of 1200 plants of Ambon, was published posthumously in 1741. Willem ten Rhijne, who collected information in Japan and was mentioned earlier, also belonged to this genre. These men gathered, assimilated, and translated culturally specific knowledge into a written, general format recognizable to a European readership, thereby adding to the model or tradition of plant description that can be traced back to Dioscorides' *De Materia Medica* (c. 40–90 CE) and Pliny the Elder's *Naturalis Historia* (c.23–79 CE). The HNB and the HM became additions to this traditional European knowledge system.

The HNB is a posthumous work. Georg Marcgraf died in Angola around 1644, four years before the publication of the treatise. Before departing to Africa, he had entrusted Count Johan Maurits with his studies and notes about Brazil. Moreover, when he passed away, he left two chests containing a book, an Arabian dictionary, a herbarium, various manuscripts (two of which in Portuguese), natural history manuscripts and drawings, an astronomical manuscript, zoological specimens, seeds, dried roots, fruits, and insects.[50] Apart from the insects, which were sold in an auction in Haarlem, the remaining items were distributed between the University of Leiden and a few fellow naturalists. One of them was Johannes de Laet, who had been corresponding with Marcgraf about Brazilian natural history and therefore was given his colleague's natural history drawings and manuscripts, as well as his herbarium.[51] In this way, he became responsible for editing what would soon become the HNB. In fact, Johan Maurits personally commissioned the edition and publication of the treatise; his patronage is clearly indicated in the title page of the tome.[52]

De Laet, born in Antwerp to a well-to-do Protestant merchant family, was indeed the ideal choice for this task. Having studied theology and classic languages, this Leiden scholar had already published books about philology, theology, geography, and history by the time he was given Marcgraf's papers. Likewise, he had translated and edited works of natural history (such as Pliny the Elder's *Naturalis Historia*) and had a profound personal interest on this subject. In 1625, he had published *Nieuwe Wereldt ofte Beschrijvinghe van West-Indien*, which could serve as a guide for colonization projects in the Americas.[53] Furthermore, he was one of the founders of the WIC and served as one of its directors. As such, he had access to a wealth of information about the Americas and could make sure that the HNB would be published in a lavishly illustrated and well-prepared edition.

Organizing Marcgraf's notes was certainly a challenge, for the naturalist had written them in ciphered code, fearful that someone might steal them.[54] Apparently, he trusted the key to the codes to Johan Maurits, who in turn gave it to De Laet. In the preface to the HNB, De Laet explains to the reader that the manuscripts were "imperfect and disordered" when given to him.[55] In addition to decoding the ciphers and transcribing Marcgraf's and Piso's notes, he also wrote and added more than a hundred notes, mostly about botany. As to the illustrations, he included some missing images of plants, which he supposedly drew according to dried specimens left by Marcgraf in his herbarium. Other plants to which De Laet had no direct access were sent to him by fellow naturalists. In fact, De Laet was one of the nodal points in a network of scholars that built an intense and lively circuit of knowledge exchange in the early modern Low Countries.[56] According to Eric Jorink, "De Laet generously shared the information and artefacts with his fellow scholars and collectors. His collection was the basis for his publications and speculations. Objects, drawings, descriptions and inscriptions were constantly related to the classical and contemporary literature."[57] Consequently, De Laet's contribution in the HNB cannot be ignored: his textual and visual additions, as well as the way he organized the chapters, are a critical part of what this treatise became and the impact it had for centuries. For instance, both the HNB and the HM became some of the main sources of information for Swedish botanist Carl Linnaeus (1707–1778), who used the information on the tropical flora of Malabar and Brazil, along with several other books published in Europe about plants in Asia, Africa, and Latin America, in his 1735 *Systema Naturae*, that soon became the standard system for the classification of plants.

Professional artists, draftsman, and people who specialized in printing made a living in the rapidly expanding printing industry in Amsterdam and Leiden. Both the HM and the HNB benefitted from this expertise. In the case of the HM, although sketches of the plants, seeds, and flowers in the books and the enlarged details were from Malabar, as well as all information on their medicinal usage, they were incorporated into the format of a European book. The design and sketch on the title page, as well as the style of preface and introduction in the HNB and the HM, were European. The system by which some plants and animals were classified together or put in different categories was also European. The South Asian input, therefore, was limited to names of plants in Malayalam, Arabic, and other local languages and to information on their medicinal usage. The content of the HM was thus essentially South Asian but the structure in which it was organized for publication was European, i.e., based on Pliny's classification order. Elements of medico-botanical knowledge with illustrations from Malabar based on Ayurvedic epistemologies and precepts had been incorporated into an entirely European format and structure of knowledge. A similar argument can be made for the HNB, whose title page became a model for other works on natural history, including Piso's 1658 *De Indiae utriusque*.

The context in which the two treatises were produced partly in Latin America and South Asia and partly in Europe is by no means an exhaustive explanation to the creation, collection, and circulation of knowledge. For example, it does not do justice to the many layers of historical and intellectual experiences that helped shape these treatises. Nor does it take into consideration the heavy contribution of Indigenous experts and local populations in the making of these treatises. Recently, art historian Amy Buono has introduced the concept of "catalogue" to understand the HNB, a treatise – she argues – which aspires to present a systematic listing of its subject and "presents a selection, a curation, of Brazil's plants and animals and orders them according to the simple taxonomy of plants, fish, insects, etc."[58] This description is also apt for the HM. In the next section, we advance the argument and propose that the HNB and HM are catalogues of knowledge on the natural world, read and presented through the lens of early modern scholarship.

Locating Knowledge in the HNB and HM

Publishing the HNB and HM entailed the transformation of local information from the northeastern coastal regions of Brazil and Malabar respectively into a format that could reassert universal validity.[59] It is remarkable, however, that the people of South Asia and Brazil did not receive the final product of knowledge about their milieu. South Asian knowledge, handed down over generations mostly in oral form and some in written form, was lost. Early modern Malabar held some parts of its knowledge in Malayalam and others in Sanskrit. There were also continual links to the classical Sanskrit learning and to which additions were made from time to time creating "systemic knowledge." The palm leaf *olas* of Itty Achuthan are an example of Malabar's systematic documentation of botanical knowledge and medicinal practices, but their content is lost forever. Even if the HM provides evidence of local knowledge, it does not conserve it beyond information. From the HM, the Malabar traditions of classification cannot be discerned, they were probably not considered worthy of preserving. In contrast, oral knowledge and knowledge practices of Indigenous peoples and of the descendants of enslaved African peoples continued to be passed onwards in oral traditions in Suriname and Brazil and in non-systematized forms, as recent work by ethnobotanist Tinde van Andel and others has been pointing out.[60] More research is needed, however, to understand the relationship between seventeenth-century Tupi and African knowledges as presented in the HNB and contemporary practices in South America.

The making of the HNB and HM demonstrates combined efforts of several agents – American, African, Asian, and European – in transforming knowledge to the European system. A complex network of agents allowed for the final production of the treatises: Brahmins, vaidyas, kings and chiefs, savants, merchants, and craftsmen of Malabar, and similarly in South America, local informants who spoke different Tupi or Gê ("Tapuia") languages and culturally belonged to different ethnic groups, as well as (the descendants of) enslaved Africans in Brazil, and, in the Low Countries, university professors, editors, publishers, and savants. Yet it was the combination of infrastructures such as universities, publishing houses, and the social milieu where individuals could socially and economically benefit from producing books in Europe that eventually led to the centralization of knowledge in Europe. In Europe, knowledge had become a commodity and there was a market for books. In Malabar, access to knowledge was a privilege of a select few – the Brahmins. In Indigenous Brazil, access to knowledge and the ability or right to put knowledge into practice was directly related to the different social organizations of the Tupi and other Indigenous groups, for instance, dividing practical knowledge into male/female tasks or restricting medical and ritualistic practices to shamans.

Some historians have brought forward the role of brokers or middlemen who operated in the "contact zones" transferring information and knowledge between cultural and linguistic barriers.[61] It has been proposed that the new knowledge was thus "hybrid" knowledge, as it took the best of both; yet one ought not to forget that trade in pharmacological substances existed in the Indian Ocean and Latin American and Caribbean region long before the arrival of the Portuguese and the Dutch.[62] The European contribution to the structure of trade and transfer of knowledge was to link them to

Europe by sea and bring together information from Asia, West Africa, and the Americas to Europe, and subsequently publish it in printed format.

The HNB and HM merged information from Latin America and South Asia with that from Europe. They declared their own hybridity. No names of individual Indigenous informants are stated in the HNB but clearly their contribution was substantial. Writing a little bit later, Van Reede clearly attributes the information in the HM to Itty Achuthan, Vinayaka Pandit, Appu Bhatt, and Ranga Bhatt, and even published their letters in the first volume. In the third volume, he explained the process of how he had accumulated all the knowledge from Malabar. The HM shows how knowledge moved from one intellectual and practical setting to another and how some aspects of it were preserved, while others were lost. The same can be read in the HNB, despite the fact that the names of Indigenous and African informants and, many times, their ethnic identities remain uncertain.

Convergence of Multiple Knowledge Systems

Collecting and compiling oral knowledge about the natural world is a universal phenomenon. The many different peoples of Brazil and South Asia were active collectors of knowledge. They had their own systems of classification and their own ways of understanding and registering the natural world. When the Europeans arrived in these regions, a whole new process of intercultural knowledge production and circulation emerged. Both in Brazil and South Asia, it is within the Portuguese empire that such knowledge-making processes were first put in motion. As Timothy Walker has shown, Jesuits played a major role in collecting medical knowledge from South American Indigenous groups and disseminating that information throughout the Portuguese colonial empire.[63] Jesuits relied on local informants, often the same local people they were trying to convert as part of their missionary work, to explain the use of South American drugs, as well as native healing methods. Subsequently, that information was compiled and described in manuscript accounts and letters that circulated through the Portuguese colonial empire from South America to Europe, Africa, and Asia.

However, while the information collected and spread within the Portuguese empire remained largely limited (in access) to the Portuguese empire itself, as it was considered "strategic imperial commercial information," one century later, the Dutch were particularly fast in publishing their findings and making it accessible to broad audiences.[64] The making of both the HNB and the HM demonstrates very similar and comparable processes in place. Because of the emergent print and visual culture in the seventeenth-century Low Countries, these two exemplary treatises on the natural world got published in beautifully illustrated multiple volumes, which appealed to European readerships. It is not fortuitous that the

HNB was put together and published within five years after Johan Maurits' return from Brazil to Europe, as Johan Maurits had a clear political agenda of self-aggrandizement. The publication of the HM took much longer due to the death of editors, lack of funding, and eventually also the sudden death of Van Reede. Yet, the HM volumes continued to be published posthumously and even had multiple editions and a Dutch translation. Van Reede and other editors also gained social capital through the publication of the HM.

These remarkable treatises had multiple locations of production. For the case of Brazil, historiography has argued that book publishing during the colonial period was incidental, and remained centered on religious books and materials used for the conversion of Indigenous peoples.[65] It was only after the arrival of the Portuguese royal family, escaping from the Napoleonic wars in Europe in the early nineteenth century, that libraries and book printing gained momentum in Brazil. Book publishing for profit was also unknown in South Asia. There was little social and economic incentive to write down all the information available and publish a book on the local flora in Malabar. Neither the head of the state, the Raja of Cochin, nor the Brahmins and vaidyas, nor laymen had any motivation to compile information under one title and make it available through publication. On the contrary, social status was gained by maintaining knowledge within the family, furthering it first by mastering the classical texts and then putting forward new thoughts and proofs through the delicate art of debate, persuasion, and skillful convincing. While Johan Maurits and Van Reede gained from the process of publication of the treatises, the case in Brazil and Malabar was different.

There is little documentation of knowledge flowing from the Low Countries toward South Asia and Latin America and then becoming popular at this early stage of interaction, especially about the natural world. At least in the sixteenth, seventeenth, and eighteenth centuries, in the Low Countries knowledge remained accessible because of book printing. The fact that most such books were in Latin helped in cross-cultural reading among Europeans.[66] Individuals like Johan Maurits, Piso, and Van Reede, and many of their contemporaries, used this industry to articulate new knowledge gathered from overseas and in the process gained social status and wealth.

Scholarly engagement with local knowledge and local informants did not end with collecting information. On the contrary, the process came full circle when that same knowledge once reached Europe and needed to be confirmed, expanded, or tested. Examples from published correspondence show that early modern scholars relied on their "New" World networks to ask and inquire, from Indigenous and local peoples and others with first-hand experience in South America, whether certain pieces of information were useful and reliable. Authentication or confirmation of information was therefore an essential part of early modern knowledge-making practices. That was the case for scholars in the Low Countries and in Britain alike. Henry Oldenburg, secretary of the London Royal Society, was responsible for requesting objects and information from all parts of the world. Via contacts in Lisbon, he reached Portuguese America and, in August 1671, wrote a long letter to a Jesuit in Bahia with numerous questions about the air, water, earth, medicinal flora, and fauna in Brazil.[67] A. Rupert Hall and Marie Boas Hall have identified the HNB as one of the main sources of information from which Oldenburg derived his detailed questionnaire.[68] The extended use of Tupi names to describe the different plants and animals is remarkable in this letter. In the same manner, Oldenburg frequently asks if Piso's reports of Indigenous Brazilians using certain plants and animals for medicinal purposes were correct: "do the more villainous Brazilians hang the aforesaid toads in the sun, collect their bile and foamy spittle, and keep these for their more secret and slow-acting poisons?"[69] Whether or not this letter was ever answered is thus far unknown. Likewise, the physical presence of Indigenous people in Europe was well-known to scholars as they reached out to them as a source of gathering and expanding information on the natural world in the Americas. In a letter from London physician Dr. Charles Goodall to the philosopher John Locke, dated July 1687, Goodall thanks Locke for having sent him samples leaves of the quinquina tree (*Cinchona pubescensc*, native to the Andes and a source of quinine). He proposes a series of questions about the growth of this particular tree, as well as of its uses by the people of South America to be asked "to any of the natives of Peru brought into Europe, or any Spaniards, who were born and lived some time in that Countrey: To any Merchants, who have long traded at Lima, Quito or other parts of Peru: To any Priests who have Spent some years there."[70]

Surely much more research is needed to establish the perceived differences in epistemologies and views of the natural world between and within Europe, Latin America, and South Asia. Judging from the case of the HNB and HM, Latin American and South Asian attitudes to nature were not very different from those of Europe, in terms of the awareness of a necessity to classify plants and collect information on their growth, usage, and medicinal values. The idea of an epistemological difference was conceived by scholars of a later period, writing with a nationalistic zeal. Intellectual trends of information gathering and knowledge production, a characteristic of the early modern world, were without any doubt a global phenomenon.

This shows that a process of inquiring, gathering, and verifying information about the natural world took place with an aim to draw from them organized and taxonomized information, which is referred to as knowledge. In other words, other than a center-periphery binary of information gathering (in Brazil and South Asia) and knowledge creation (in Europe) in the early modern period, knowledge of the natural world was continuously constructed in multiple locations concomitantly, which converged in print form in Europe.

Notes

1 This chapter was researched by Anjana Singh during a postdoctoral fellowship at the London School of Economics: *Useful and Reliable Knowledge in Global Histories of Material Progress in the East and the West (URKEW)*, which has received funding from the ERC grant agreement no. 230326. Mariana Françozo's research for this chapter is part of the ERC project *BRASILIAE. Indigenous Knowledge in the Making of Science*, directed by Dr. Mariana Françozo at Leiden University and funded by the European Research Council Horizon 2020 Research and Innovation Programme (Agreement No. 715423). Unless otherwise indicated, all translations in this chapter are our own.

2 Andre Gunder Frank, *ReORIENT: Global Economy in the Asian Age* (Berkeley, Los Angeles, CA, and London, UK: University of California Press, 1998).

3 Kapil Raj, *Relocating Modern Science: Circulation and the Construction of Knowledge in South Asia and Europe, 1650–1900* (Basingstoke, UK: Palgrave Macmillan, 2007).

4 Harold J. Cook, *Matters of Exchange: Commerce, Medicine, and Science in the Dutch Golden Age* (New Haven, CT: Yale University Press, 2007); Jürgen Renn, ed., *The Globalization of Knowledge in History: Based on the 97th Dahlem Workshop*, Max Planck Research Library for the History and Development of Knowledge Studies 1 (Berlin: Edition Open Access, 2012); Dániel Margócsy, *Commercial Visions: Science, Trade, and Visual Culture in the Dutch Golden Age* (London, UK: University of Chicago Press, 2014).

5 Kapil Raj, "Thinking Without the Scientific Revolution: Global Interactions and the Construction of Knowledge," *Journal of Early Modern History* 21, no. 5 (2017): 445–458, doi: 10.1163/15700658-12342572.

6 Susanne Friedrich, Arndt Brendecke, and Stefan Ehrenpreis, *Transformations of Knowledge in Dutch Expansion*, Pluralisierung & Autorität 44 (Berlin: De Gruyter, 2015); Siegfried Huigen, Jan L. de Jong, and Elmer Kolfin, *The Dutch Trading Companies as Knowledge Networks* (Leiden: Brill, 2010).

7 Londa Schiebinger, *Plants and Empire: Colonial Bioprospecting in the Atlantic World* (Cambridge: Harvard University Press, 2004).

8 Christopher Bayly, *Empire and Information: Intelligence Gathering and Social Communication in India, 1780–1870* (Cambridge, UK: Cambridge University Press, 1996), 3–4.

9 Richard Grove, *Green Imperialism: Colonial Expansion, Tropical Island Edens and the Origins of Environmentalism, 1600–1860* (Cambridge, UK: Cambridge University Press, 1995), passim and especially 79.

10 Michiel van Groesen, "Abraham Willaerts: Marine Painter of Dutch Brazil and the Atlantic World," *Oud Holland: Journal for Art of the Low Countries* 132, no. 2–3 (2019): 65–78, doi: 10.1163/18750176-1320203002.

11 Ernst van den Boogaart and Rebecca Parker Brienen, eds., *Informações do Ceará de George Marcgarf (Junho-Agosto de 1639)* (Rio de Janeiro: Index/Petrobrás, 2002), 9–10.

12 Van den Boogaart and Brienen, *Informações do Ceará*, 9–10.

13 Neil Safier, "Beyond Brazilian Nature: The Editorial Itineraries of Marcgraf and Piso's *Historia Naturalis Brasiliae*," in *The Legacy of Dutch Brazil*, ed. Michiel Van Groesen (New York, NY and Cambridge, UK: Cambridge University Press, 2014), 175.

14 Júnia Ferreira Furtado, "Tropical Empiricism: Making Medical Knowledge in Colonial Brazil," in *Science and Empire in the Atlantic World*, ed. James Delbourgo and Nicholas Dew (London, UK: Routledge, 2008), 136.

15 Henrique Carneiro, "O Saber Indígena e os Naturalistas Europeus," *Revista Trajetos* 7, no. 13 (2009), 55.
16 Timothy Walker, "The Medicines Trade in the Portuguese Atlantic World: Acquisition and Dissemination of Healing Knowledge from Brazil (c.1580–1800)," *Social History of Medicine* 26, no. 3 (2013): 403–431, doi: 10.1093/shm/hkt010, 407.
17 Willem Piso and Georg Marcgraf, *Historia Naturalis Brasiliae: In qua non tantum Plantae et Animalia, sed et Indigenarum Morbi, Ingenia et Mores Describuntur et Iconibus supra Quingentas Illustrantur* (Leiden and Amsterdam: Elzevier, 1648), 25–26. See also Georg Marcgraf, *História Natural do Brasil* (São Paulo: Imprensa Oficial, 1942 [1648]), 25–26.
18 Safier, "Beyond Brazilian Nature," 177.
19 Marcgraf, *História*, 92.
20 Ibid, 4.
21 Ibid, 123.
22 Ibid, 17.
23 Ibid, 39.
24 Ibid, 94–95.
25 See Marcgraf, *História*, 203–205 for the feathers.
26 Ibid, 21.
27 Ibid, 39.
28 Ibid, 140.
29 Mariana Françozo, *De Olinda a Holanda: O Gabinete de Curiosidades de Nassau* (Campinas: Ed. Unicamp, 2014), 204–205; Virginie Spenlé, "'Savagery' and 'Civilization': Dutch Brazil in the Kunst- and Wunderkammer," *Journal of Historians of Netherlandish Art* 3, no. 2 (2011), doi: 10.5092/jhna.2011.3.2.3.
30 Marcgraf, *História*, 192, 227–230.
31 Francisco Ximenes, *Quatro Libros de la Naturaleza, y Virtudes de las Plantas, y Animales* (Mexico: en casa de la viuda de Diego Lopez Daualos, 1615).
32 Marcgraf, *História*, 38.
33 Ibid, 40.
34 Ibid, 61.
35 Ibid, 61–62.
36 Ibid, 93.
37 Ibid, 68.
38 See Marcgraf, *História*, 108.
39 Information on his life and works are available in the scholarship of Hugo K. s'Jacob and Johannes Heniger. Johannes Heniger, *Hendrik Adriaan van Reede tot Drakenstein (1636–1691) and Hortus Malabaricus: A Contribution to the History of Dutch Colonial Botany* (Rotterdam: A. A. Balkema, 1986). Heniger gives a complete picture of the life of Van Reede and the making of the *Hortus Malabaricus*. Unless otherwise stated, all information on Van Reede has been collected from this biography.
40 M. Antoinette P. Roelofsz, *De Vestiging Der Nederlanders Ter Kuste Malabar* (Leiden: Brill, 1943).
41 Hugo K. s'Jacob, *The Rajas of Cochin, 1663–1720: Kings, Chiefs and the Dutch East India Company* (New Delhi: Munshiram Manoharlal, 2000).
42 Pius Malekandathil, *Portuguese Cochin and the Maritime Trade of India: 1500–1663* (New Delhi: Manohar, 2001).
43 Anjana Singh, *Fort Cochin in Kerala, 1750–1830: The Social Condition of a Dutch Community in an Indian Milieu*, Tanap Monographs on the History of Asian-European Interaction, V. 13 (Leiden: Brill, 2010).
44 Bayly, *Empire and Information*, 13.

45 Anjana Singh, "Botanical Knowledge in Early Modern Malabar and the Netherlands: A Review of Van Reede's *Hortus Malabaricus*," in *Transformations of Knowledge in Dutch Expansion*, Pluralisierung & Autorität 44, Susanne Friedrich, Arndt Brendecke, and Stefan Ehrnpreis (Berlin: De Gruyter, 2015), 187–208.

46 *De Lijkstatie van Baron van Rheede te Suratte*, 1693, engraving, 165×267 mm. Inscription states: Burial Procession of Baron Hendrik Adriaan van Reede, Lord of Mijdrecht, died on 15 December 1691 and buried in January 1692.

47 Also written as *Malabaarsche Cruythof* and other versions of the spelling. Hendrik Adriaan van Reede tot Drakenstein, *Malabaarse Kruidhof, vervattende het Raarste Slag van Allerlei Soort van Planten die in het Koningrijk van Malabar Worden Gevonden: Nevens der selver Blommen, Vruchten en Saden* (Amsterdam: de weduwe van Joannes van Someren, de erfgenamen van Jan van Dyck, Henrik Boom en de weduwe van Dirk Boom, 1689). The title of the Dutch edition translates to "Malabar's Garden of Spices: Consisting of the rarest types of all sorts of plants that were found in the kingdom of Malabar, including their flowers, fruits, and seeds. The book was published in Amsterdam by the widow of Johannes van Someren, the inheritors of Jan van Dyck, Henrik Boom and the widow of Dirk Boom."

48 W. Wijnaendts van Resandt, *De Gezaghebbers der Oost-Indische Compagnie op hare Buiten-comptoiren in Azië* (Amsterdam: Uitgeverij Liebaert, 1944), 182.

49 See K.S. Manilal, ed., *Van Reede's Hortus Malabaricus English Edition: With Annotations and Modern Botanical Nomenclature*, 12 vols. (Thiruvananthapuram: University of Kerala, 2003). K.S. Manilal, ed., *Van Reede's Hortus Malabaricus Malayalam Edition: With Annotations and Modern Botanical Nomenclature*, 12 vols. (Thiruvananthapuram: University of Kerala, 2008)

50 Peter Whitehead, "Georg Marcgraf and Brazilian Zoology," in *Johan Maurits van Nassau-Siegen 1604–1679: Essays on the Occasion of the Tercentenary of his Death*, ed. Ernst van den Boogaart (The Hague: The Johan Maurits van Nassau Stichting, 1979), 432–433.

51 For a discussion of De Laet's use of Marcgraf's herbarium, as well as its trajectory and use in Denmark where it is presently kept, see Peter Wagner, "Das Markgraf-Herbarium," in *Sein Feld war die Welt: Johan Moritz von Nassau-Siegen (1604–1679)*, eds. Gerhard Brunn and Cornelius Neutsch (Münster: Waxmann, 2008), 233–245. For a detailed account of the judicial battle for Marcgraf's inheritance between Marcgraf's brother Christian and Leiden professor Jacob Gools, see Th. J. Meijer, "De Omstreden Nalatenschap van een Avontuurlijk Geleerde," *Jaarboekje voor Geschiedenis en Oudheidkunde van Leiden en Omstreken* 64 (1972), 63–76. Meijer argues that Marcgraf's herbarium was given to Jacob Gool, who later gifted it to Adolph Vorstius.

52 Johan Maurits of Nassau-Siegen had already commissioned another book in this same time period: the *Rerum per Octennium in Brasiliae*, a panegyric history of the eight years of his governorship in Brazil, written by Dutch humanist Caspar Barleus (1584–1648) and published in Amsterdam in 1647. Caspar Barleus, *Rerum per Octennium in Brasilia et alibi nuper Gestarum sub Praefectura Illustrissimi Comitis I. Mavritii, Nassoviae* (Amsterdam: Joan Blaeu, 1647).

53 For De Laet's earlier work, see Johannes de Laet, *Nieuvve Wereldt ofte Beschrijvinghe van West-Indien, wt Veelderhande Schriften ende Aen-teeckeninghen van Verscheyden Natien by een Versamelt* (Leiden: Isaack Elzevier, 1625). The *Nieuwe Wereldt* had four editions: the first one, in Dutch, appeared in 1625; a second one, also in Dutch, in 1630; a third one, translated to Latin, in 1633; and, finally, an edition in French in 1640. Each new edition presented new, updated information about the "New" World.

54   Specifically, he seems to have been afraid that Piso would take his notes and publish them as if they were his own. See Whitehead, "Georg Marcgraf," 434 and Rebecca Parker Brienen, "Georg Marcgraf (1610–c.1644): A German Cartographer, Astronomer and Naturalist-Illustrator in Colonial Dutch Brazil," *Itinerario* 25, no. 1 (2001): 85–122, doi:10.1017/S0165115300005581, 93.

55   De Laet in Marcgrave, *História,* no page number.

56   Sven Dupré and Christoph Lüthy, eds., *Silent Messengers: The Circulation of Material Objects of Knowledge in the Early Modern Low Countries,* Low Countries Studies on the Circulation of Natural Knowledge 1 (Berlin: LIT, 2011).

57   Eric Jorink, "Noah's Ark Restored (and Wrecked): Dutch Collectors, Natural History and the Problem of Biblical Exegesis," in Dupré and Lüthy, *Silent Messengers,* 173.

58   Amy Buono, "Interpretative Ingredients: Formulating Art and Natural History in Early Modern Brazil," *Journal of Art Historiography* 10, December (2014), 1–21, doi: 10.17613/M6610VR5Z, 6–7.

59   Grove, *Green Imperialism,* passim.

60   Tinde van Andel, "The Reinvention of Household Medicine by Enslaved Africans in Suriname," *Social History of Medicine* 29, no. 4 (2016), 676–694, doi: 10.1093/shm/hkv014.

61   Simon Schaffer, Lissa Roberts, Kapil Raj, and James Delbourgo, *The Brokered World: Go-Betweens and Global Intelligence, 1770–1820,* Uppsala Studies in History of Science 35 (Sagamore Beach, MA: Science History Publications, 2009); and Cook, *Matters of Exchange.*

62   Anna Winterbottom, *Hybrid Knowledge in the Early East India Company World,* Cambridge Imperial and Post-Colonial Studies (London, UK: Palgrave Macmillan, 2016). See also Anna Winterbottom, "Company Culture: Information, Scholarship, and the East India Company Settlements 1660–1720s" (PhD diss., Queen Mary University of London, 2010). Panchanan Maheshwari and Kapil Raj, "A Short History of Botany in India," *Journal of the University of Gauhati* 9 (1958), 3–32.

63   Walker, "The Medicines Trade," 404, 411–412.

64   Walker, "The Medicines Trade," 428. See also Junia Ferreira Furtado, "Tropical Empiricism," 132. Recent studies have been pointing to the importance of Iberian naturalists and Iberian networks of scholars and connoisseurs in the Americas. They guaranteed an enormous influx of information on the fauna and flora of the "New" World that would eventually reach other parts of Europe, even if not always in the format of published books. See Maria Luz López Terrada, "Flora and the Hapsburg Crown: Clusius, Spain, and American Natural History," in Dupré and Lüthy, *Silent Messengers,* 43–68.

65   Lawrence Hallewell, *O Livro no Brasil: Sua História* (São Paulo: Edusp, 2005 [1982]). See also Aníbal Bragança and Márcia Abreu, eds., *Impresso no Brasil: Dois Séculos de Livros Brasileiros* (São Paulo: Editora Unesp; Rio de Janeiro: Fundação Biblioteca Nacional, 2011) and Luiz Carlos Villalta, "Montesquieu's Persian Letters and Reading Practices in the Luso-Brazilian World (1750–1802)" in *Enlightened Reform in Southern Europe and Its Atlantic Colonies, c. 1750–1830,* ed. Gabriel Paquette (London, UK and New York, NY: Routledge, 2009), 119–141.

66   Cook, *Matters of Exchange.*

67   Henry Oldenburg, "Inquiries for Brazil" in *The Correspondence of Henry Oldenburg,* volume VIII, eds. Alfred Rupert Hall and Marie Boas Hall (Madison, WI: The University of Wisconsin Press, 1971), 220–248 (letter 1780). This letter was a reply to an earlier correspondence sent to Oldenburg by Thomas Hill in Lisbon, where he offers the services of "… a Jesuit, son of a Dutch man, who is very curious, ingenious and inquisitive man, and especially desirous to serve the

R.S ....." and, having travelled all over Brazil, was able to respond to any questions Oldenburg may have. Thomas Hill to Henry Oldenburg, 13 July 1671, in Hall and Hall, *The Correspondence of Henry Oldenburg*, 155 (letter 1747). See also Dominik Collet, "Big Sciences, Open Networks, and Global Collecting in Early Modern Museums," in *Geographies of Science*, eds. Peter Meusburger, David Livingstone, and Heike Jons (New York, NY: Springer, 2010), 124.

68 Hall and Hall, *The Correspondence of Henry Oldenburg*, 248.
69 Oldenburg, "Inquiries for Brazil," 247.
70 Edward S. de Beer, ed., *The Correspondence of John Locke*, volume 3 (Oxford, UK: Clarendon Press, 1989), 233.

2 Portuguese Parallels

Comparing Analogous Efforts toward Codifying Indigenous Medicinal Knowledge in Seventeenth- and Eighteenth-Century Brazil[1]

Timothy D. Walker

Introduction

Historians of medicine have largely undervalued, or failed to appreciate entirely, the importance of Portuguese global exploration and colonial settlement, especially relating to botanical prospecting, the treatment of tropical diseases, and the circulation of non-European Indigenous medical substances. The Portuguese maritime empire came into being long before that of any European rival, and was far more diverse geographically, culturally, and ecologically. Thus, Portuguese exposure to a broad spectrum of Indigenous healing ideas lasted far longer than that of any other seafaring imperial nation, and was key to conveying valuable regional medical information, not just back to Europe but around the globe.

This chapter will present and explicate rare information regarding circumstances and techniques for the assimilation and codification of medicinal knowledge from Brazil during the seventeenth and eighteenth centuries. Through the use of Portuguese colonial medical texts (translated excerpts), and images from those texts, this chapter will provide insight into how Portuguese colonial agents gathered information – in response to healthcare exigencies – for the systematic codification of healing plant knowledge, and for the application of Indigenous medicinal substances. These reflections will, in turn, allow for a substantive comparison of Portuguese colonial efforts that are analogous to the well-known Dutch medicobotanical publication of the mid-seventeenth century, Piso and Marcgraf's *Historia Naturalis Brasiliae* (henceforth HNB). Considered comparatively, we can gain some insight into differences in how Dutch and Portuguese colonial agents approached the matter of codifying Indigenous medicinal knowledge in seventeenth- and eighteenth-century Brazil.

The Portuguese crown and colonial administrators in Lisbon (the Conselho Ultramarino) took an interest in discovering new medicinal plants from their colonies in the seventeenth century, but this was especially so beginning in the late eighteenth century. Sustained high losses

DOI: 10.4324/9781003362920-3

of human capital in the tropics – not only among European soldiers, administrators, and settlers but also among valuable enslaved persons in the Atlantic and Indian Ocean regions – prompted this curiosity. Crown authorities, desperate to find effective remedies that could reduce casualties, revived their interest in discovering new Indigenous remedies and commissioned medical authorities to catalogue native plants that could be of therapeutic and commercial use to support Portuguese imperial endeavors.

Most of the information presented will be excerpted from little-known seventeenth- and eighteenth-century Portuguese primary sources: pharmacological and botanical compilations (medical and apothecary guides) in manuscript, compiled in colonial Brazil, beginning at about the time of Piso and Marcgraf's work, the HNB, and continuing into the next century.

Colonial Dutch Medical Works about Brazil: Piso and Marcgraf's *Historia Naturalis Brasiliae* (Leiden and Amsterdam: Elzevier, 1648)

Competitive Dutch and Portuguese colonial ambitions collided in Brazil (among other imperial locations) in the seventeenth century, with significant ramifications for European botanical and medical knowledge. Dutch attempts to conquer and occupy parts of the Brazilian coastline (1624–1654) in the hope of exploiting the colony's rich sugar-producing regions also resulted in a new treatise of ethno-pharmacology being written in the Netherlands. The HNB, published in Leiden and Amsterdam in 1648, was the work of three authors, all trained physicians: Georg Marcgraf; Willem Piso, who had been the personal physician of Johan Maurits of Nassau-Siegen (governor of Dutch Brazil, to whom the work is dedicated; he was the grandnephew of stadtholder William of Orange); and Johannes de Laet. With its beautiful, technically accurate engraved images of plant and animal life, the HNB represents the first comprehensive, encyclopedic, and widely circulated natural history of South America.[2] Their work stands to this day as the most exhaustive description ever published of the flora and fauna of the Dutch-occupied regions of Brazil.[3]

Willem Piso in particular distinguished himself as an investigative medical pioneer and became one of the earliest northern European authorities on tropical medicine from South America. He contributed four "books" (chapters) with medical themes to the HNB, contained in a section entitled *De Medicina Brasiliense*. These chapters include detailed descriptions of endemic regional diseases and the plant-based remedies Indigenous peoples used to treat them. Piso wrote one discrete section of the work devoted entirely to the spectrum of Brazilian medicinal plants, and another focusing on venoms, venomous animals, and native antidotes.[4]

A decade later, in 1658, Piso published a new edition of the HNB under his name alone. The revised version ran to 14 volumes and included dramatically expanded observations on Brazilian remedies. Piso introduced Protestant Europe to many of the unique Brazilian medicinal plants (ipecacuanha, copaiba, jalapa) and Indigenous healing methods long known to Iberian missionaries and settlers in South America. Piso's achievement far outshone the Portuguese for detail, clarity, and scientific method.[5]

What explains this asymmetry in quality of empirical methodology and achievement between works produced by authors from the Low Countries and those from Lusophone Iberia? Due to a number of cultural and geographic factors in the realm of natural philosophy, contemporary Portuguese and Dutch approaches to "botanizing" diverged in significant ways.

By the mid-seventeenth century, learned gentlemen in the intellectual centers of the Dutch Republic had been deeply influenced by currents of reason and empiricism that then flowed through Protestant northern Europe. The Netherlands had emerged as a center of experimentation – a hub in the scientific revolution. For example, in the leading commercial center of Middelburg, capital of the province of Zeeland, craftsmen like Zacharias Janssen, who fashioned lenses for spectacles, claimed to have built a workable compound microscope by 1600 (whether entirely of his own invention or by working off the ideas of others is unclear).[6] A few years later, his neighbor and rival spectacle maker Hans Lippershey created a practical telescope, for which he applied for a patent in 1608.[7] Such innovative optical tools opened new avenues for investigation into the cosmos, large and small – and allowed for stunning realizations about the natural world that quickly followed.[8] The Dutch explorers who entered Brazil during the first half of the seventeenth century thus came primed with capacities and a worldview that their brethren from Portugal lacked.[9]

Portuguese investigators whose personal cosmologies were conditioned far to the south, without direct access to the new instruments of the scientific revolution, did not benefit from the strong intellectual impetus such inventions helped to create in Holland. The Portuguese were thus geographically and philosophically well outside the ebb and flow of these epistemological tides. Iberian researchers prior to the nineteenth century therefore lacked both the scientific instruments and the conceptual tools to approach Brazilian flora and fauna with anything like the precision that their Dutch counterparts were capable of a century and a half prior.[10]

Ironically, though the HNB significantly furthered the dissemination of plant remedies and healing knowledge from Brazil to northern Europe, its success as a seminal publication of natural philosophy had little impact on continental Portugal. Because it was the work of a Protestant empiricist physician from a nation with which the Portuguese were at war, Piso's masterpiece was banned by the Inquisition Board of Censorship and so could not circulate freely in Lusophone lands. Holy Office *familiares* and royal agents boarded inbound ships and inspected cargoes from northern Europe to insure that the HNB and other prohibited texts would not be imported to Portugal or Brazil.[11] Hence, Piso's work remained virtually unknown in the Portuguese-speaking world until the later nineteenth century.[12]

That said, in an additional ironic twist, it is possible that among the learned and cosmopolitan Jesuit missionaries who traveled and read the latest natural philosophy publications outside of Portuguese-controlled areas, the HNB was not only known but also occasionally used as a reference tool. For example, a Spanish Jesuit and naturalist, Friar Pedro de Montenegro, who embarked on a mission to Paraguay in approximately 1693, where he studied the medicinal plants of South America, clearly knew of Marcgraf and Piso's work. Working in Paraguay between 1710 and 1711, Montenegro completed a manuscript focused on the trees and plants of the Tucuman region and Brazil. He wrote in Spanish, primarily, but employed Guarani and Tupi names for the plants featured throughout his text. He also revealed explicitly having seen the HNB, which he referred to repeatedly throughout the botanical manuscript he composed, citing from it for comparative material.[13]

This irony is further extended when one appreciates just how much early Portuguese maritime exploration – and the publications resulting therefrom that focused on tropical flora, fauna, and medicine – had inspired the composition of the HNB. When Johan Maurits was preparing his invasion of Brazil in 1636, he actively recruited natural philosophers to accompany his Dutch expedition, specifically to have trained personnel on hand to exploit the abundance of natural resources he knew was available in the Portuguese American territories.[14] One of these men, Johannes de Laet, had earned his degree at Leiden University, where he was a contemporary of the pioneering botanist and physician Carolus Clusius, perhaps the most influential of all sixteenth-century European horticulturists. Clusius founded the experimental botanical gardens adjacent to the university grounds at Leiden; these would have been familiar to Johannes de Laet.

More importantly, working in collaboration with the renowned publishing house in Antwerp operated by Christophe Plantin and Jan Moretus, Carolus Clusius had overseen the translation, publication, and dissemination of some of the most important seminal texts regarding newly discovered plants and medicines to emerge from the beginnings of European overseas expansion. These works included a volume about the medicines of India by a Portuguese *converso* physician, Garcia da Orta (originally published in Portuguese in 1563; translated into Latin in 1567); a book about the medicines of the West Indies by the Spanish physician Nicolás Monardes (published in Spanish in 1565; translated into Latin in 1574); and a treatise on medicinal substances of the East Indies that expanded and corrected Garcia da Orta's work by African-born Portuguese physician Cristovão da Costa (published in Spanish in 1578; excerpted in Latin translation and included in Clusius' illustrated compendium *Exoticorum Libri Decem*, published in 1605).[15] Willem Piso, Georg Marcgraf, and

Johannes de Laet would have been intimately familiar with all these texts, which effectively served as archetypal models for their compilation of the HNB.[16]

Seventeenth- and Eighteenth-Century Lusophone Colonial Medical Works about Brazil

In the course of the seventeenth and eighteenth centuries, either well before the advent of the Dutch or long after their withdrawal, several secular Portuguese medical practitioners attempted to codify descriptions of useful healing substances found in "the Brazils," classifying them according to their respective provenances and applications.[17] By the mid-eighteenth century, a broad range of South American medical substances had entered common pharmaceutical usage in continental Portugal.[18]

Indigenous medicinal plants that Portuguese settlers adopted and exported from Brazil in significant quantities beginning in the sixteenth century included derivatives of *cacau* (medicinal chocolate and cocoa butter, the latter used to treat skin ailments); *ipecacuanha* (also called *cipó*), a reliable emetic and diaphoretic; cinchona bark (also called *quina* or *quineira*), arguably the most important remedy found in the "New" World, essential to treating malaria and other tropical fevers; *jalapa*, an effective purgative; *copaiba*, to treat gonorrhea; and *salsaparilha*, administered against syphilis and skin diseases.[19] More than any others, these Brazilian remedies circulated in the Atlantic World medicines trade, becoming commercially and medically significant, and achieving widespread usage elsewhere in the Portuguese empire. Further, in the HNB, Piso and Marcgraf had provided especially detailed descriptions of these plants, as well as the healing properties for which they were known. Their meticulous, precise treatment of Brazilian flora was superior in many ways, not only to those descriptions and illustrations produced by earlier Portuguese colonial authors but also to later ones in the eighteenth century, discussed below.

Other plant-derived drugs originating with Indigenous practices in Brazil but found inculcated into the Portuguese medical lexicon elsewhere in the Lusophone world included *abutua* root (drunk in a decoction to treat fever, or as a purgative and stimulant), *tacamahaca* gum (a bitter resin used as a topical balm), *guaiaco* gum (used to treat wounds or sores or mixed in a beverage drunk to ease sore throat pain), *mechoacão* (a white *jalapa* root), and *almécega* gum (a tree resin chewed to relieve pain).[20] Brazilian healers employed *maracujá* (passion fruit) juice to treat fevers, pineapple juice (*ananás*) to dissolve kidney stones, cashew fruit juice (*cajú*) for fever and stomach ailments, and inga fruit (*ingá*) for addressing liver problems.[21] All of these substances were standard, commonly stocked medicines in Brazilian apothecary shops in the seventeenth and eighteenth centuries, and they also could be found in continental Portuguese pharmacies (mainly in cosmopolitan seaports like Lisbon and Porto, but also at larger communities in the interior like Coimbra and Évora).

Indigenous peoples of Brazil thus made important contributions to "Western" medicine during the early modern period, but typically did so anonymously and indirectly through European intermediaries, who often failed to discuss the original human sources for this knowledge. Though they gathered ethno-botanical information systematically, European colonists frequently ignored the rationale for, or altered the application of, Indigenous healing techniques to meet their own ends and exigencies. As the pioneering Swiss anthropologist and ethnographer Alfred Métraux pointed out, though early writers on Brazil listed many medicinal plant substances, they rarely indicated whether Indigenous peoples had traditionally used them for healing purposes or if it was the Europeans who discovered their virtues as remedies.[22] Brazilian historian of medicine Júnia Ferreira Furtado describes this process with clear succinctness, as follows: "empiricism in the colonial context in fact became a way to extract *practical* knowledge from natives without embracing their heathen and superstitious beliefs about nature, magic, and their gods. By cementing this new knowledge in written language, Europeans converted it into an erudite framework. They adopted practical medical techniques from the Indigenous populations, but they insisted on divorcing these techniques from the natives' superstitious accounts of why they functioned the way they did. This process was emblematic of the European colonists' appropriation of technical knowledge while separating these techniques from the native systems of thought in which they initially appeared."[23]

One example of how information about Indigenous Brazilian medicine traveled through cross-cultural exchanges and by word of mouth in the Portuguese colonial world can be found in the well-known *Diálogos das Grandezas do Brasil*, an early seventeenth-century manuscript attributed to the Lisbon-born *converso* planter and writer Ambrósio Fernandes Brandão (1555–1618).[24] This laudatory text, meant to promote colonization in Brazil by describing the region's plants, animals, people, and political circumstances, is structured as a set of conversational exchanges between two imaginary colonists, one an experienced Brazil hand and the other an apprehensive newcomer. The manuscript's six dialogues systematically dismiss purported dangers or problems popularly attributed to the South American colony and then proceed to extol its vast potential. The second of the dialogues addresses climate and disease directly: Brandônio (the experienced protagonist) reluctantly affirms that sickness does exist in Brazil, but forcefully asserts that the diseases are so "mild and easy to cure that they almost don't deserve the name."[25] He touts the healing value of tobacco, salsaparilha, and pajamarióba,[26] all medicinal plants in Indigenous use, among the many local "leaves and juices of herbs" that can be used to treat common ailments.

Brandônio next refers to a recipe for a purgative derived from a pine nut that is roasted inside a guava fruit. Wounded soldiers, he claims, are readily healed with native "copaúba" (perhaps the same as copaiba, described above) or a balsam confected from plants found in the southern provinces. A major advantage of the colony, he says, is that illnesses that usually prove fatal in India can be cured in Brazil because of the unique native medicines available there.[27] The neophyte Alviano asks dubiously if the Portuguese actually make frequent use of native remedies; Brandônio assures him that the European colonists do, finding the local materia medica so effective that surgeons, barbers, and bloodletters from the *metropôle* are rarely called upon for their services.[28] Although Brandão's attempt to attract settlers to Brazil downplayed health risks in the territory and exaggerated the probable efficacy of native plants, his description of colonial remedies derived from Indigenous medical techniques was nevertheless accurate, drawn from practical experience. Brandão also referred specifically to Indigenous medicines being used, already at the beginning of the seventeenth century, in attempts to cure not only Europeans in Brazil but also the enslaved African people, whose deaths he deemed to be economically damaging for the colony.[29]

After the mid-seventeenth century, the focus of the Portuguese colonial enterprise shifted from Asia to the Atlantic World. In Goa in 1563, Portuguese *converso* physician Garcia da Orta had published his seminal work on medicines from India, *Coloquios dos Simples e Drogas e Cousas Medicinais da Índia*, a book that introduced European natural philosophy to many Asian healing plants and methods.[30] The only other major original treatise concerning Indigenous medicine produced by the Portuguese during this period, *Tratado de las Siete Enfermedades*, printed in Lisbon in 1623, dealt primarily with the healing flora used in West Africa and Brazil. Written in Latin and Spanish by the Coimbra-trained doctor Aleixo de Abreu, who had spent 15 years practicing medicine in Luanda and Salvador da Bahia, it was not improved upon as a didactic tool regarding Atlantic tropical medicine in the Lusophone context until the mid-eighteenth century.[31]

Regrettably, a faltering of coordinated scientific inquiry at the institutional and state level was typical of early modern Portuguese administration in their overseas holdings through the mid-eighteenth century: while the Jesuits, the Santa Casa da Misericórdia (Holy House of Mercy), and individual *médicos* continued to operate hospitals and collect medical data in the Atlantic sphere, eastern Africa, and Asia, state resources and activities, limited as they were, remained focused on expanding trade or protecting territory. Not until the waning years of the eighteenth century did the Portuguese crown, moved at last by Enlightenment sensibilities that the rest of Europe had long since embraced, send out state-sponsored scientific expeditions to systematically study and record flora and fauna of the overseas empire.[32] At that late date, such activities cast a broad net, focusing on colonial holdings in Brazil, Africa, and coastal southwest India.[33]

Enlightenment Medicine and the Portuguese Empire: Increasingly Systematic and Comprehensive Guides about Brazilian Medical Substances

The century that was drawing to a close in contemporary Brazil had been literally a golden age; extraordinary wealth derived from the colony's booming gold and diamond mines drew many Portuguese immigrants, including a surprisingly meager handful of trained physicians and surgeons from the metropole.[34] There they confronted an entirely unfamiliar disease environment of the Amazon and South American tropics, as well as an Indigenous healing culture shaped by the context of the unique flora and fauna at their disposal, blended with influences to healing methodologies resulting from the massive regular annual influx of enslaved African peoples. Those western *médicos* with inquisitive spirits produced handbooks, papers, or guides to the novel Indigenous healing plants they encountered.[35] Some with a more scholarly bent went much further, creating detailed works for publication, apparently in the hope of reaching a wide audience back in Europe among interested researchers in medicine, botany, or natural philosophy.

In Bahia, a Portuguese-born physician, Francisco António de Sampaio, resident in the district of the erstwhile colonial capital city, Salvador, undertook an ambitious project of pharmacological botany, which he entitled *History of the Vegetable, Animal and Mineral Kingdoms, Pertaining to Medicine*. His venture – a medical field guidebook broad in its scope and objective – was a particularly incisive undertaking for someone who was essentially a country doctor in Brazil in the late eighteenth century. Sampaio compiled his two-volume work between 1782 and 1789 at Cachoeira, the main agricultural market town situated at the highest navigable point on the Paraguaçú River in the fertile Bahian hinterland around the Bay of All Saints. Fair copies of each volume, hand-written and bound together with stunning original painted illustrations of Brazil's flora and fauna, are deposited in the special collections division of the Biblioteca Nacional in Rio de Janeiro.[36]

Sampaio embraced the role of an Enlightenment-era *médico*, who clearly wanted to expose his countrymen to a deeper knowledge and understanding of the traditional Brazilian Indigenous medicinal plants with which he regularly worked. Indeed, the project, because of its structure and parameters, shows telltale signs of having been produced by commission, most likely at the behest of colonial authorities in Lisbon or possibly Bahia. The two extant manuscript tomes each contain highly detailed descriptions of a variety of native South American plants, a summary of their healing virtues, proper doses to administer to patients, and methods for applying each remedy to the sick.[37]

Francisco António de Sampaio had been born at Vila Real in northern Portugal but immigrated to Brazil as a young man. Where he completed his medical studies is not clear, but he never enrolled in any formal course at the Coimbra University faculty of medicine.[38] Most likely he trained as an apprentice with a licensed physician, or through a residency at the Todos-os-Santos Hospital in Lisbon, before leaving Portugal. In 1762, King José I granted him a license as a physician, with the right to practice surgery.[39] In Brazil, he became an approved surgeon with a license to practice medicine granted by the Bahian colonial senate.[40] Sampaio then held the post of surgeon at the Hospital of São João de Deus in Cachoeira for nearly two decades. In the 1780s and 1790s, Sampaio corresponded with officials of the new Academia Real das Ciências de Lisboa, and was registered among its members in 1798; his letters reveal that he submitted copies of his work for consideration and inclusion in the academy's library, and that he engaged with a network of like-minded medical botanizers within the Portuguese Atlantic.[41]

Volume I of his work, completed in 1782, described medicinal plants in 219 manuscript pages, supported by another 20 pages of color miniature paintings that skillfully rendered many of the plants described in the text. Although it is highly improbable that Sampaio had read the HNB, nevertheless there are some organizational similarities between the two texts. The plants described in Sampaio's first volume are organized into 12 sections according to their contemporary medicinal applications. European usage in the colony generally mirrored practices gleaned from Indigenous peoples, but also reflected innovative applications pioneered by the colonizers through their own experimentation. Groups of plants evincing astringent, anti-venom, anti-colic, anti-spasmodic, purgative, and anti-venereal healing qualities are each treated in their own discrete chapters. Sampaio provided practical information about how to identify, gather, and preserve each plant discussed, together with instructions about various therapeutic applications, proper medical preparation, elaborated recipes, and dosages.

To provide a stronger sense of *History of the Vegetable, Animal and Mineral Kingdoms, Pertaining to Medicine*, consider a few examples of medicinal preparations drawn from Sampaio's text:

- "Cuyaté." This plant is listed in the section on "resolvent" medicines. Cuyaté is made from the fruit produced by a small tree; Sampaio reports that it has a very bitter juice. The unripe fruit was to be roasted among hot coals; then the fruit was opened and the hot "meat" (pulp) of the fruit was applied topically, directly to tumors or swellings in the body. This application could be repeated as many times as the sufferer found necessary.[42] Curiously, in the HNB, Marcgraf explained how Tupi peoples opened the "Cuiete" fruit, cleaned it, and then hollowed it out to make a drinking vessel from the calabash.[43]

- "Cajazeira." A strongly astringent plant with a fruit that emits a rich aromatic smell, this healing remedy was indicated for diarrhea, internal hemorrhaging, and "other illnesses for which astringents are used." The strongest effects were achieved by using the interior part of the bark of a young cajazeira tree. An infusion of the plant could be drunk, or applied externally as needed to any affected areas of the body.[44]

- "Gameleira." Listed among purgatives and emetics, this remedy was composed of the milky sap (liquor) of the tree, extracted by making incisions in the bark of the tree "during nights when the moon is waning." According to Sampaio, gameleira was one of the most potent vegetable purgatives found in Brazil; he used a few ounces of it to treat patients suffering from dropsy (edema) and cachexia (wasting illness). The sap could also be dried in the sun to form a gum or powder; this substance could then be made into a bolus and swallowed to achieve the same effect. However, when taken in its condensed form, Sampaio warned that the purgative effects of gameleira were exceptionally pronounced and could be dangerous.[45]

- "Dandá." Listed among plants that work as an anti-venom or febrifuge, Sampaio reports that the "sagacious inquiries" of the Brazilian Indigenous had never found a better remedy for the bite of snakes and other poisonous creatures. Dandá was a type of reed abundant on riverbanks in the Bay of All Saints hinterland. The medicine was made from the plant root, which could be pulled up and applied when fresh, or alternatively dried. In either case, as soon as one was bitten, the root was to be masticated, the resulting juice swallowed, and the masticated pulp applied as a poultice over the affected part of the body.[46]

While the plant name indicates a probable west African origin, this is most likely an example of enslaved Africans adopting and incorporating Brazilian Indigenous peoples' traditional medicines into their own healing repertoire. Due to the horrific conditions endured during the trans-Atlantic trade in enslaved persons' middle passage, it is unlikely that Africans succeeded in the widespread transplanting of much of their own materia medica from their homeland to the Americas – though it happened in some circumstances. Recent research by ethnobotanists suggests that enslaved African people did manage to take some types of seeds of medicinal and foodstuff plants with them to the "New" World.[47] However, enslaved African people most frequently sought South American plants that were similar to familiar African ones – with an analogous medical application – and employed them instead, sometimes replicating healing rituals or compounds remembered

from home, thus creating a new, hybridized healing culture in demograph-
ically diverse colonial Brazil.[48]

- "Caróba." Included among anti-venereal curative plants, the medicinal leaves of caróba evidenced no smell but had an unpleasant taste;
  Sampaio touted this as one of the most effective plants ever discovered
  for such sexually transmitted maladies. According to the text, caróba
  found growing by the sea is more effective than plants harvested in the
  interior. It could be prepared either as an infusion and drunk, or the
  solution could be used to wash venereal sores. The leaves could also be
  powdered and mixed with other simples to create a paste. The paste
  was to be spread topically on venereal sores, including syphilitic tumors, scrofula, or other eruptions (boubas).[49]

To aid in plant identification for neophytes to Brazil, Sampaio included an
alphabetical index of each plant name, carefully noting their Indigenous
Tupi and Guarani names, as well; this is one of the most extraordinary
components of the manuscript. Sampaio's intent was not to be culturally
sensitive, of course; his objective was practicality. In a colonial context
that relied heavily on non-Portuguese speaking Indigenous Brazilians to
help source medicinal plants from the undeveloped interior, a phonetic
guide to useful plant names was an essential addition to his pharmacopoeial handbook. He also commented knowingly on the use and efficacy
of non-native medicinal plants, like coffee, pepper, and cinnamon, that
the Portuguese had introduced from Europe or their overseas territories
in Asia.[50]

   Although the painstaking, protracted work was obviously intended for
publication to a broad readership in the trans-Atlantic medical and scientific community, for unknown reasons the project never went beyond the
manuscript stage. Sampaio's work may have been considered too provincial, or dated, or he simply may have been unable to win the support of
an influential patron in the metropole. Had the work been published, it
would have contributed greatly to knowledge of regional Brazilian healing plants in the Lusophone world (though, without an edition translated
into French or English, its significance outside the Portuguese market
would have been small in the late eighteenth century). Instead, it languished as an unprinted manuscript until the late twentieth century. In
1969, both volumes of Francisco António de Sampaio's *História dos
Reinos Vegetal, Animal e Mineral* were published together in Rio de
Janeiro as a special issue of the journal *Anais da Biblioteca Nacional*.
Despite his affiliation with the Academia Real das Ciências, circulation
of Sampaio's fair copy text was probably minimal; therefore, its ultimate
impact as a conduit of information, though uncertain, was likely very
meager.

Bento Bandeira de Mello: Untitled Memorandum about Medicinal Plants of Brazil (1788). Manuscript Held at the Arquivo Nacional de Torre do Tombo, Lisbon, Portugal

Elsewhere in Portuguese America, at about the same time, a similar state-directed effort to gather medicinal knowledge was underway – this one explicitly mandated by colonial authorities in the metropole. In 1788, Brazilian physician and natural scientist Bento Bandeira de Mello submitted a lengthy memorandum on frequently used Indigenous medicines in his home region, the coastal northeast of Brazil. De Mello was responding to a direct royal order from Queen Maria, transmitted through the Overseas Council; he had been charged with creating an alphabetical list of medicinal plants, fruits, and roots from the territories of Pernambuco and Paraíba, with commentary concerning their curative effects.[51] His annotated roster, containing 59 different South American medicinal plants, runs to 24 manuscript folios, now archived at the Portuguese National Archive (Torre do Tombo) in Lisbon. To date it has never been published; the document remains little-known, even among Lusophone historians of medicine.

Examples of native healing plants de Mello discussed in his compilation include various types of ipecacuanha (also called cipó), a reliable emetic and diaphoretic; cinchona bark (also called quina or quineira), a febrifuge used to treat malaria and other endemic fevers throughout the Portuguese network of colonies; jalapa, an effective purgative; copaiba, the bark and plant oil of which was used internally and externally to treat gonorrhea; and salsaparilha, administered against syphilis and skin diseases.[52] More than any others, these particular Brazilian remedies had attracted a broader market within the Atlantic medicinal economy, gaining widespread medical usage not only in South America but elsewhere in the Portuguese empire. Demand for these plants grew steadily until they became significant commodities, both medically and commercially. By the end of the eighteenth century, these Indigenous Brazilian healing products, in bulk or in prepared remedies, commonly could be found in ships' seaborne medical chests or stocking Portuguese pharmacy shelves from Lisbon to Mozambique, Goa, Timor, and Macau.[53]

Per royal instructions, de Mello sent specimens of many of these plants to the royal botanical garden of the Ajuda Palace in Lisbon, where they were assessed for their medical usefulness, as well as for their suitability for transplant to other imperial regions.[54] Hence, the impact of his work carried much farther than the palace chambers of the Conselho Ultramarino. The desired end of this official initiative, of course, was to further Portuguese aims by reducing chronic, unacceptably high wastage of human resources through injury and illness.

To better understand the nature of Bento Bandeira de Mello's pharmacopoeial text, consider a few examples of descriptions of lesser known

Brazilian medicinal plants, all of which he presented as being in popular contemporary use[55]:

- "Angelico do Mato." Effects that it produces: the grated or sliced root of this plant, when boiled with water, was a laxative and efficacious against high fevers; the same mixture could be drunk daily to control bouts of epilepsy (*gota coral*). This mixture, when blended with egg whites, was used to treat bloody dysentery and, used topically, it facilitated the healing of skin lesions or shingles, making them less virulent. For women, drinking an infusion made of the boiled root with sugar was believed to help release the fetus during childbirth or the placenta thereafter; however, for the same reason, "Angelico do Mato" may also have been used in higher concentrations as an abortifacient.[56]
- "Barbatimão." Effects that it produces: the husk or bark of this plant was crushed and applied directly to heal any type of wound or skin abrasion. Such wounds could also be washed with an infusion made from the same bark. The same infusion was used by women following childbirth; according to the text, they washed themselves with a strong barbatimão solution to tighten and rejuvenate their genitalia, to make them appear "as if they had never before known a man." This highly astringent plant infusion, when concentrated in a reduction, also served as a cure for bloody dysentery, as well as a treatment for deep tissue *corrupção* – an infection, like gangrene.[57]
- "Jabarandim." Effects that it produces: the leaves and roots of this plant, when boiled in water and then swished around in the mouth, were thought to be good for toothache, and, when drunk, it is indicated for snakebite, or irritations of the skin caused by "other poisonous creatures." The root could also be crushed into a paste and placed directly on the affected gums and teeth, or on the snakebite or irritated skin. The roots could also be chewed to make the required paste; if swallowed, it is reported to promote urination. Grated bark scrapings, and the juice from the leaves, were said to desiccate and cure venereal sores. The plant could also be ground into a paste and used as a plaster externally to treat intermittent chills from colds or fevers.[58]
- "Malicia de Mulher." Effects that it produces: the pulverized root of this plant, literally called "women's malice," was part of a remedy for sharp pains and general aches, especially of the chest and trunk of the body. The leaves, crushed and mixed with saliva, would also strip calluses from the feet. Its juice, or an infusion of the root, mixed with sugar, could be drunk before sleeping to expel kidney stones; the same herb mixed with sweet olive oil and applied topically helped cure an inflamed liver or spleen or back pain. The plant's juice when mixed with honey treated canker sores. The leaves of "Malicia de Mulher"

were known to be poisonous, for which the same plant's root served as an antidote.[59]

- "Quandú." Effects that it produces: an infusion of quandú when drunk daily was thought to alleviate or even cure the symptoms of tuberculosis. The plant's seeds could be boiled unsalted; the resulting broth and seeds could then be consumed to create the same anti-tubercular effect. Sprouts of this plant's seeds could be chewed raw to stop bleeding in the mouth, or the masticated seedling paste could be rolled in spider webs and placed on wounds as a bandage to staunch bleeding. The juice of such quandú seedlings, when mixed in water and gargled, reduced swollen or inflamed glands in the throat. A quandú leaf infusion also treated halitosis. Further, ashes from the incinerated leaves were applied to skin burns to heal such wounds without scarring.[60]

- "Timbauba." Effects that it produces: its grated bark was mixed with a small amount of water in the early evening and left out overnight to soak and evaporate. The resulting paste was applied to male patients with gonorrhea, or to other types of venereal pustules. According to the text, this treatment could be repeated up to six times. This remedy also could be applied to women with the same venereal diseases, or women whose natural menstruation cycle had been disrupted, or those who suffered from hemorrhoids.[61]

Conclusion

Over three centuries of Portuguese colonial rule, Brazil's Indigenous peoples and unique flora contributed significantly to the eclectic, cosmopolitan, syncretic medical culture that developed within the colonial Lusophone world. Systematic Portuguese maritime exploration in the tropics began in the 1430s, predating by far any other European effort; consequently, Portuguese exposure to tropical diseases, as well as various Indigenous cultural methods of treating them, lasted far longer than that of any other European nation. Necessity combined with long familiarity resulted in the marked Portuguese tendency for receptiveness toward the adoption and dissemination of Indigenous medical practices. Through numerous ecclesiastical, commercial, and medical channels, knowledge of South American botanicals and healing techniques circulated throughout the Atlantic World and beyond, enriching medical resources in European imperial enclaves around the globe. Some drugs, like cinchona, actually acted as a catalyst, allowing for the marked expansion of European colonial power in the tropics during the nineteenth century into territories where they had previously been denied sustained access, due to a chronic inability to treat the endemic fevers prevalent there.

As we have seen, exchanges of medical knowledge in colonial Brazil occurred on a variety of levels; in any given case, much depended on the preexisting knowledge, skills, and requirements of the persons directly involved. In the missionary context, protracted exchanges were often substantially more complex – and intellectually more profound – than those rapid transactions conducted between sick Portuguese soldiers, *bandeirante* explorers deep in the bush, a harried colonial provincial official, or even a ship's or regimental surgeon, and the native Tupi or Guarani shamans with whom they interacted. Like their martial or mercantile coreligionists, Jesuit priests and lay missionaries often relied on Indigenous cures to treat their own tropical maladies contracted in the service of the Church; however, their greater patience and investment of time for evangelizing ends resulted in a more subtle and detailed understanding. The theological implications of their reliance on Indigenous materia medica, though, must have given pause to evangelicals who were constantly at pains to demonstrate the superiority of "Old" World religion and culture.[62]

Only in the late eighteenth century, near the end of colonial era in Brazil, did an interest in the exploitation of Indigenous remedies awaken within the core government administrators of the Portuguese empire, leading to the commissioning of systematic surveys of Brazil's medicinal plants by professionals with botanical or medical training. The result, though, was little different from what Jesuit priests had produced over 200 years before: lists of known Indigenous plants, compiled meticulously but unscientifically, combined with descriptions of their use according to how colonists and Indigenous people applied them. By this time, centrally organized scientific exploration had long been underway in rival Dutch, English, and French colonies; in each respective imperial enterprise, empiricism had become a common component.[63] Indeed, the HNB may be seen as a very early example of such deliberate national scientific endeavors, supported or even planned and carried out with a broad imperial purpose in mind. Even though Jesuit missionaries in Portuguese colonies had conducted the earliest European medical and ethno-botanical explorations in the tropics, the broad historic impact of later Portuguese scientific endeavor was limited by having begun belatedly, and because little of these efforts ever achieved, or were even meant for, circulation within Europe's international scientific or intellectual communities. Instead, these newly gathered reports, the knowledge within them still regarded as strategic imperial commercial information, usually remained hidden from the general public and therefore languished, awaiting discovery by modern researchers.

Curiously, much of this medical and botanical information from disparate geographic origins would over time come to be gathered together for practical application in the mid-Atlantic Ocean, on the islands of Terceira and São Miguel, part of the Azores archipelago. This may seem like an unlikely location for a nexus of empirical information about medicinal flora, until one considers that the Azores had become, due to natural oceanic wind and current patterns, a crossroads of trade – and hence a crossroads of some very specialized knowledge. After all, during the early modern period, the Kingdom of Portugal had built up a diverse colonial and commercial maritime trading network throughout the Atlantic world, incorporating biologically rich colonies extending from those along the length of West Africa to the Atlantic archipelagos of Madeira and Cape Verde to, most notably, Brazil in South America.[64] Collectively, the key ports of the Azores Islands – Angra do Heroísmo on Terceira, Horta on Faial, and Ponta Delgada on São Miguel – would emerge as mid-ocean havens of primary importance during the age of sail, serving variously as victualing points, harbors of refuge during storms, and repair centers for merchant and military ships returning from Asia, Africa, and the Americas.[65]

It should come as no surprise, then, that diverse plant-based remedies from all these regions became well known in the Azores; mariners carried plants there, where many thrived as transplants to the temperate island climate. At the Military Hospital of Nossa Senhora da Boa Nova, established in the seventeenth century at the regional capital, Angra do Heroísmo, experimental tropical medicines brought from the Portuguese colonies became regular components of the treatment regimens prescribed by the staff. The surviving documentation of this medical facility, covering the period 1766–1820, is extensive and detailed; it includes daily prescription and medical recipe lists compiled by the chief surgeon, chief physician, and *enfermeiros* (nurses).[66] The Dutch, however, with priorities elsewhere and kept at bay by massive harbor fortifications at Angra do Heroísmo, were never privy to the medical knowledge that accumulated over time on these islands.

The 1648 publication of the HNB marks a significant milestone in the development of botanical and zoological science, and the way natural philosophers gathered information about the flora and fauna of any given region. The work had a long and powerful influence among botanists and other natural philosophers regarding the systematic codification of plant and animal knowledge, including the pioneering Swedish physician Carl Linnaeus (1707–1778), for whom it was a guiding example for the system he developed in the eighteenth century for the organization and classification of species. His system of naming organisms, called "binomial nomenclature" and first fully articulated in the tenth edition of his *Systema Naturae* (published 1758), became the standard method of taxonomy adopted by modern science. Linnaeus had become acquainted with the HNB while he was a student at Uppsala University; later, his formalized universal nomenclature system was definitively inspired by, and evolved in part from, the tentative patterns laid down in Piso and Marcgraf's earlier work on Brazil.[67]

So, by considering comparatively analogous Dutch and Portuguese written works of the seventeenth and eighteenth centuries that focused on describing the flora and fauna of Brazil, often with a view toward understanding their novel medicinal benefits and applications, we have achieved some insight into the significant differences between their respective approaches to codifying colonial Indigenous medicinal knowledge. Portuguese efforts over time were pragmatic and situational, often determined by unique and highly varied localized conditions found across their South American possessions and shaped by the overriding need to maintain the health and labor effectiveness of their very limited but diverse human resources. That is, Portuguese exigencies focused on safeguarding the relatively few non-native personnel of their colonial population, which were spread over a very broad geographic area. As a result, their overall efforts may be characterized as being chronologically protracted, and revealing an intimate familiarity with the natural environment and Indigenous people, but ultimately uncoordinated, incremental, and fragmented.

By comparison, the Dutch, though their access to Brazil over a period of just three decades was relatively brief and was focused on a much more limited geographic area, approached the codification of South American flora, fauna, and Indigenous healing knowledge in a much more systematic and comprehensive way, with a goal of universal coverage and inclusion of the natural resources available in their occupied zone. Hence, the HNB emerges as a consolidated Dutch achievement, driven by an ambitious empirical vision. Further, the Dutch effort may be characterized as being the production of a far more "modern" scientific publication that is markedly superior to the piecemeal mosaic of texts that the Portuguese produced over two and a half centuries of colonial rule. In this regard, the HNB serves as an eloquent example of an advanced northern European work of knowledge production that clearly outstrips, in terms of sophistication of methodology and technical execution, analogous contemporary Portuguese efforts.

Notes

1 This work was supported by funding from the U.S. National Endowment for the Humanities, the American Institute for Indian Studies, the University of California Davis Department of Nutrition, the University of Massachusetts Dartmouth Office of the Provost and Center for Portuguese Studies and Culture, the Fundação Calouste Gulbenkian, and the Fundação Luso-Americana para o Desenvolvimento. In addition, for logistical support in Brazil, thanks are due to the Fundação Oswaldo Cruz and the Biblioteca Nacional do Rio de Janeiro. Unless otherwise indicated, all translations in this chapter are my own.
2 David Freedberg, "Ciência, Comércio e Arte," in *O Brasil e os Holandeses*, ed. Paulo Herkenhoff (Rio de Janeiro: Sextante Artes, 1999), 195–202.
3 Júnia Ferreira Furtado, "Tropical Empiricism: Making Medical Knowledge in Colonial Brazil," in *Science and Empire in the Atlantic World,* ed. James Delbourgo and Nicholas Dew (London, UK: Routledge, 2008), 137.
4 Freedberg, "Ciência, Comércio e Arte," 200–201.
5 Furtado, "Tropical Empiricism," 132–138.

6   Albert van Helden, Sven Dupré, and Rob van Gent, *The Origins of the Telescope (History of Science and Scholarship in the Netherlands)* (Amsterdam: Amsterdam University Press, 2010), 43.

7   van Helden, Dupré, and van Gent, *Origins of the Telescope*, 32–36.

8   James Delbourgo and Nicholas Dew, "Introduction: The Far Side of the Ocean," in Delbourgo and Dew, *Science and Empire*, 8.

9   Harold J. Cook, *Matters of Exchange: Commerce, Medicine, and Science in the Dutch Golden Age* (New Haven, CT: Yale University Press, 2007), 267–268, 289–292.

10 Francisco Bethencourt, "Enlightened Reform in Portugal and Brazil," in *Enlightened Reform in Southern Europe and its Atlantic Colonies, c.1750–1830*, ed. Gabriel Paquette (London, UK: Ashgate Press, 2009), 41–44.

11 José Sebastião Silva Dias, "Portugal e a Cultura Europeia: Séculos XVI a XVIII," *Biblos* 28 (1952): 292–297. The Inquisition in Portugal was not abolished until 1821; government censorship of scientific texts continued until the implementation of a new liberal constitution the following year.

12 Licurgo de Castro Santos Filho, *História da Medicina no Brazil, do Século XVI ao Século XIX*, 2 vols. (São Paulo: Editora Brasiliense Limitada, 1947), vol. 2, 43–48.

13 John Carter Brown Library, Manuscript collection, Codex SP 36; ff. 161, 220, 224–225. I am grateful to Dr. Mariana Françozo of Leiden University for alerting me to this reference.

14 Furtado, "Tropical Empiricism," 132–133.

15 Garcia da Orta, *Colóquios dos Simples e Drogas he Cousas Medicinais da Índia* [...] (Goa: João de Endem, Colégio de São Paulo, 1563); Garcia da Orta, *Aromatum et Simplicium aliquot Medicamentorum apud Indos Nascentium Historia* [...] (Antwerp: Christophe Plantin, 1567); Nicolás Monardes, *Dos Libros, el Uno que Trata de Todas las Cosas que se Traen de Nuestras Indias Occidentales, que Sirven al Uso de la Medicina, y el Otro que Trata de la Piedra Bezaar, y de la Yerva Escuerçonera* (Seville: Hernando Diaz, 1565); Carolus Clusius, *De Simplicibus Medicamentis ex Occidentali India delatis quorum in Medicina Usus Est* (Antwerp: Christophe Plantin, 1574); Cristóvão da Costa, *Tractado de las Drogas y Medicinas de las Indias Orientales* (Burgos: Martin de Victoria, 1578); Carolus Clusius, *Exoticorum Libri Decem, quibus Animalium, Plantarum, Aromatum, aliorumque Peregrinorum Fructuum Historiae Describuntur* (Leiden: Officina Plantiniana Raphelengii, 1605).

16 Neil Safier, "Beyond Brazilian Nature: The Editorial Itineraries of Marcgraf and Piso's *Historia Naturalis Brasiliae*," in *The Legacy of Dutch Brazil*, ed. Michiel Van Groesen (New York, NY and Cambridge, UK: Cambridge University Press, 2014), 169–177.

17 Aleixo de Abreu, *Tratado de las Siete Enfermedades, de la Inflamacion Vniuersal del Higado [...] de la Terciana y Febre Maligna, y Passion Hipocondriaca [...] del Mal de Loanda, del Guzano y de las Fuentes y Sedales [...]* (Lisbon: Pedro Craesbeeck, 1623); Luís Gomes Ferreira, *Erário Mineral* (Lisbon: Officina de Miguel Rodrigues, 1735); Biblioteca Nacional do Rio de Janeiro (BNRJ), Manuscritos, I-12,01,019: Francisco Arsenio de Sampaio, *História dos Reinos Vegetal, Animal e Mineral*, vol. I (1782) and vol. II (1789).

18 José Pedro Sousa Dias and Rui Pita, "A Botica de S. Vicente e a Farmácia nos Mosteiros e Conventos da Lisboa Setecentista," in *A Botica de São Vicente de Fora* (Lisbon: Associação Nacional das Farmácias, 1994), 12–20.

19 Saul Jarcho, *Quinine's Predecessor: Francesco Torti and the Early History of Cinchona* (Baltimore, MD: Johns Hopkins University Press, 1993), 102–104, 297–298; Andreas-Holger Maehle, *Drugs on Trial: Experimental Pharmacology*

and Therapeutic Innovation in the Eighteenth Century (Amsterdam: Rodopi Press, 1999), 223–233; Timothy D. Walker, "The Medicines Trade in the Portuguese Atlantic World: Acquisition and Dissemination of Healing Knowledge from Brazil (c. 1580–1800)," *Social History of Medicine* 26, no. 3 (2013): 403–431, doi: 10.1093/shm/hkt010.

20 José Pedro Sousa Dias, "Inovação Técnica e Sociedade na Farmácia da Lisboa Setecentista" (PhD diss., Universidade de Lisboa, 1991), 697; Arquivo Nacional da Torre do Tombo (ANTT), Lisbon, Ministério do Reino, cx. 555, mç. 444, Bento Bandeira de Mello manuscript (1788). I am indebted to Bruno Barreiros of Universidade Nova de Lisboa for alerting me to this document.

21 Flávio Coelho Edler, *Boticas & Pharmacias: Uma História Ilustrada da Farmácia no Brasil* (Rio de Janeiro: Casa da Palavra, 2006), 26, 76–77.

22 Alfred Métraux, "The Tupinambá," in *Handbook of South American Indians*, volume 3, ed. Julian H. Steward (Washington, DC: Smithsonian Institution, 1941), 130–131.

23 Furtado, "Tropical Empiricism," 137; italics in the original.

24 Ambrósio Fernandes Brandão (attributed), *Dialogues of the Great Things of Brazil*, trans. and annot. Frederick Arthur Holden Hall, William F. Harrison, and Dorothy Winters Welker (Albuquerque, NM: University of New Mexico Press, 1987), 9–10.

25 Ambrósio Fernandes Brandão (attributed), *Diálogos das Grandezas do Brasil*, ed. José António Gonçalves de Mello (Recife: Imprensa Universitária, 1966), 49.

26 Brandão, *Dialogues,* 106, 126, note 67, Coffee senna (Linnaeus: *Cassia occidentalis*).

27 Brandão, *Diálogos,* 36–57.

28 Ibid. I am grateful to Dr. Thomas Rogers of Emory University for alerting me to these references.

29 Brandão, *Dialogues,* vii, 107–113. Two original manuscripts of this text, each with slight variations, are held in the Biblioteca Nacional de Lisboa, Portugal, and at Leiden University Libraries, The Netherlands.

30 Orta, *Colóquios dos Simples*; Garcia da Orta, *Colóquios dos Simples e Drogas he Cousas Medicinais da Índia [...]* Facsimile edition (Lisbon: Academia das Ciências de Lisboa, 1963).

31 Abreu, *Tratado.*

32 Anthony John Russell-Wood, *World on the Move: The Portuguese in Africa, Asia and America, 1415–1808* (Manchester: Carcanet Press, 1992), 83–85.

33 Timothy Walker, "Acquisition and Circulation of Medical Knowledge within the Portuguese Colonial Empire during the Early Modern Period," in *Science, Power and the Order of Nature in the Spanish and Portuguese Empires*, ed. Daniela Bleichmar, Kristin Huffine, Paula De Vos, and Michael Sheehan (Palo Alto, CA: Stanford University Press, 2009), 257–260, 267–268; José Pinto de Azeredo, *Ensaios Sobre Algumas Enfermidades de Angola* (Lisbon: Regia Officina Typografica, 1799).

34 Santos Filho, *História da Medicina,* 50–51; Eduardo de Castro Almeida, *Inventário dos Documentos Relativos ao Brasil Existantes no Archivo de Marinha e Ultramar de Lisboa,* vol. I: "Bahia, 1613–1762." (Rio de Janeiro: Officinas Graphicas da Biblioteca Nacional, 1913), 255–256, document 2917.

35 BNRJ, Manuscritos, I-47,19,20: "Anotações Sobre Medicina Popular," ff. 1–32; BNRJ, Manuscritos, I-47,23,5: "Botânica Médica Vulgar Brasileira: Drogas Orgânicas & Medicina Popular," ff. 1–17.

36 BNRJ, Manuscritos, I-12,01,019. The journal *Anais da Biblioteca Nacional* published both manuscript volumes together as a special issue. See Francisco António de Sampaio, "História dos Reinos Vegetal, Animal e Mineral, Pertencente à Medicina," *Anais da Biblioteca Nacional* 89 (1969).

37 Ibid.
38 Gisele C. Conceição, "Estudos De Filosofia Natural No Brasil Ao Longo Do Século XVIII," in *História & Ciência: Ciência e Poder Na Primeira Idade Global*, ed. Amélia Polónia, Fabiano Bracht, Gisele C. Conceição, and Monique Palma (Porto: Biblioteca da Faculdade de Letras da Universidade do Porto, 2016), 126.
39 ANTT, Chancelarias de Dom José I, livro 70, f. 282v. I am indebted to Gisele C. Conceição of the Universidade do Porto for this reference.
40 BNRJ, Manuscritos, I-12,01,019, volume I, f. 1.
41 Conceição, "Estudos De Filosofia," 126–129.
42 Sampaio, "História dos Reinos," 15.
43 Willem Piso and Georg Marcgraf, *Historia Naturalis Brasiliae: In qua non tantum Plantae et Animalia, sed et Indigenarum Morbi, Ingenia et Mores Describuntur et Iconibus supra Quingentas Illustrantur* (Leiden and Amsterdam: Elzevier, 1648), 123. I am grateful to Dr. Mariana Françozo of Leiden University for alerting me to this reference.
44 Sampaio, "História dos Reinos," 33.
45 Ibid, 33–34.
46 Ibid, 60–61.
47 Tinde van Andel, *Ethnobotany: Linking Traditional Plant Use to Health, History and Heritage* (Wageningen: Wageningen University 2016), 8–11; Tinde van Andel, "African Names for American Plants," *American Scientist*, 103 (2015): 268–275, doi: 10.1511/2015.115.268; Judith Carney, "Seeds of Memory: Botanical Legacies of the African Diaspora," in *African Ethnobotany in the Americas*, ed. Robert Voeks and John Rashford (New York, NY: Springer Publishers, 2013), 13–33.
48 Luís Gomes Ferreira, *Erário Mineral* (Collecção Mineiriana), 2 volumes, ed. Júnia Ferreira Furtado (Rio de Janeiro: Editora FIOCRUZ, 2002), vol. I, 319–446.
49 Sampaio, "História dos Reinos," 73–74.
50 BNRJ, Manuscritos, I-12,01,019, volume I, ff. 117–124.
51 ANTT, Ministério do Reino, cx. 555, mç. 444.
52 Jarcho, *Quinine's Predecessor*, 102–104, 297–298; Maehle, *Drugs on Trial*, 223–233; Arquivo Histórico Ultramarino (AHU), Lisbon, Manuscritos, AHU-ACL-CU-016, cx. 25, doc. 1311: Vicente Jorge Dias Cabral, "Ensaio Botanico de Algumas Plantas de Parte Interior do Piauí [...]" (1801); António José de Sousa Pinto, *Materia Medica* (Ouro Preto: Typografia de Silva, 1837), 21–31; José E. Mendes Ferrão, *A Aventura das Plantas e os Descubrimentos Portugueses* (Lisbon: Chaves Ferreira Publicações, 2005), 157–160.
53 AHU, São Tomé e Príncipe Collection; cx. 55, doc. 75; Ana Maria Amaro, *Introdução da Medicina Ocidental em Macau e as Receitas de Segredo da Botica do Colégio de São Paulo* (Macau: Instituto Cultural de Macau, 1992), 7–11; José Pedro Sousa Dias, "Documentos sobre Duas Boticas da Companhia de Jesus em Lisboa: Colégio de Santo Antão e Casa Professa de S. Roque," *Economia e Sociologia* 88, no. 8 (2009), 295–312; Historical Archive of Goa (HAG), India, Codex 7926: "Medicines Sent from the *Hospital Real* of Goa to the Fortress of Diu" (1785), f. 56r/v; HAG Codex 1346: "Relação dos Medicamentos que Fazem Precizo para o Hospital Publico Militar dos Ilhas de Soldar e Timor" (Dili, 5 May 1838), f. 183.
54 ANTT Ministério do Reino, cx. 555, mç. 444., f. 2.
55 All examples drawn from ANTT, Ministério do Reino, cx. 555, mç. 444 (pages not numbered).
56 Ibid.
57 Ibid.
58 Ibid.

59 Ibid.
60 Ibid.
61 Ibid.
62 António Barrera-Osorio, "Knowledge and Empiricism in the Sixteenth-Century Spanish Atlantic World," in Bleichmar et al., *Science, Power and Nature*, 219–232; Daniela Bleichmar, "A Visible and Useful Empire: Visual Culture and Colonial Natural History in the Eighteenth-Century Spanish World," in Bleichmar et al., *Science, Power and Nature*, 290–310.
63 Gabriel Paquette, ed., *Enlightened Reform in Southern Europe and its Atlantic Colonies, c. 1750–1830* (London: Ashgate Press, 2009), 1–21; Patrick Manning and Daniel Rood eds., *Global Scientific Practice in an Age of Revolutions, 1750–1850* (Pittsburgh, PA: University of Pittsburgh Press, 2016), 1–20.
64 Thomas Bentley Duncan, *Atlantic Islands: Madeira, the Azores and the Cape Verdes in Seventeenth-Century Commerce and Navigation* (Chicago, IL: University of Chicago Press, 1972), 1–6, 239–252.
65 Bentley Duncan, *Atlantic Islands*, 137–157.
66 The documentation also includes extensive internment, mortality, and financial records. Biblioteca Pública e Arquivo Regional do Angra do Heroísmo (BPA-RAH), Terceira, Azores Islands, Manuscripts, Capitánia Geral dos Açores, "Almoxarifado do Hospital Militar da Nossa Senhora da Boa Nova," 50 maços.
67 Furtado, "Tropical Empiricism," 137.

3 Cover to Cover

A Book Historical Approach to the *Historia Naturalis Brasiliae*

Alex Alsemgeest and Jeroen Bos

Introduction

When the *Historia Naturalis Brasiliae* (HNB) came from the press of a Leiden printing office in 1648, it was already the result of a complex editorial history.[1] The steps leading up to the materialization of the text arguably started in the years prior to the publication, first with field research in Brazil and later with choices made for type, composition, illustration, and coloration. After the treatise was published, the copies were distributed among booksellers, sold to customers and collectors across Europe, found their way to the European aristocracy and the civic elite, or ended up in the collections of religious orders, colleges, or universities. In the following years, decades, and centuries, as the reputation of the HNB prolonged, these copies were constantly redistributed through auctions, legacies, donations, exchanges, and confiscations. New owners not only used and interpreted the treatise in new contexts, they physically reshaped the tome by adding markings, marginalia, inscriptions, bookplates, decorations, or even new bindings to their copies. These adaptations are material evidence of dealings with the past and thus valuable sources in reconstructing the trajectory of the HNB and understanding the multiple histories of the treatise.

The HNB has been subject of research in several academic disciplines, covering historical, zoological, botanical, iconographical, and ethnographical perspectives.[2] Surprisingly, we know little about the material history of the tome. The process of printing, publication, and dissemination, as well as the initial reception, use, redistribution, and later adaptations of the treatise in new contexts, have all been studied before, but never in a comprehensive way, that is, based on multiple copies of the same treatise. There is a practical explanation for the absence of such an analysis. The study of early printed books as material objects relies heavily on the availability of copy-specific information. Even though individual libraries sometimes provide detailed information on the binding and provenance of their respective copies in local catalogues, there are very few transnational databases that offer copy-specific information as structured data. The exception arguably lies in the field of incunabula studies, with the Material Evidence in

DOI: 10.4324/9781003362920-4

Incunabula (MEI) database as the standard example of how material evidence can and should be offered as structured data.

In this contribution, we use a combination of tools from analytical bibliography and cultural history to uncover the material history of the HNB and create a framework that allows us to study its trajectory. We set up a copy census to track down as many remaining copies of the HNB as possible and record all material evidence connected to it.[3] Even though our census is only a first attempt to give a comprehensive overview of all surviving copies of the HNB, and is most certainly incomplete and imperfect, we argue that copy-specific information of more than 300 copies ensures a solid basis to further explore the social and cultural context of the HNB. In our framework, we connect the census with the model for the life cycle of the book,[4] and the concepts of heritage, cultural memory, and patrimonization, that have been used in cultural history for some time and were applied in the field of book history more recently.[5]

The model for the life cycle of the book, which traditionally describes the acts of production, publication, distribution, reception, and survival, can now be studied in relation to material evidence that is directly connected to individual copies of the HNB. The census gives a systematic overview of actors in relation to the copies at different stages in the life cycle of the book: trade networks, acquisition, transfer of ownership, reception, use, and survival of the book. The concepts heritage, cultural memory, and patrimonization help to explain how actors dealt with their copies over time and, consequently, why copies are found in one library or one country and not another, why actors brought physical changes to the book, and why books did or did not survive in certain places.

Copy Census

The copy census as a methodology has a history of its own.[6] The method derives its reputation from Sidney Lee's census of Shakespeare's first folio copies in 1902 and Seymour de Ricci's census of Caxtons a few years later.[7] The pioneers in the field obviously did an outstanding job finding, identifying, and listing dozens of copies from across the world without the aid of modern communication, electronic databases, and digital reproductions. One of the first major implications of these censuses was the understanding that numerous small technical, often coincidental interventions or errors in the production of the book lead to hundreds of small divergences in the text. Editorial changes were applied at the press to text and composition, ink was not always evenly distributed over the page, misprinted text might have been corrected halfway through the print run, type could have shifted or fallen out, and countless other things could and did happen during the process of printing. Consequently, Lee, De Ricci, and contemporaries typically focused on textual divergences and printing history and paid little attention to bindings, provenances, and marginalia. Former owners were occasionally mentioned, but not recorded in a structured way.

Understanding the history of books as a social and cultural phenomenon, researching the trajectory of specific copies, and explaining how ideas were transmitted through print is a relatively modern approach. More recent examples of copy surveys seem to focus less on the questions of textual scholarship and more on the social impact of books. Owen Gingerich's *An Annotated Census of Copernicus' De Revolutionibus* does not only list 560 known copies of two editions of *De Revolutionibus*, it also analyzes important annotations, provenances, and thus the trajectory of the copies.[8] Appendices with auction results provide insight in collectors and library development over the centuries. The approach by Gingerich has inspired other scholars to explore the ownership and trajectory of other famous books, for example Vesalius' *Fabrica*.[9]

Another recent example is the census of copies of the 1705 edition of Maria Sibylla Merian's *Metamorphosis Insectorum Surinamensium*.[10] The list was published as an appendix to a facsimile of the book, totals 67 copies worldwide, and makes an interesting comparison with our census of the HNB. First, because both publications cover natural history in roughly the same geographical area, but arguably even more so because the context wherein these books were published could not have been more different. Whereas Merian had to publish the first edition of her *Metamorphosis* as a private undertaking at her own expenses, the publication of the HNB was a highbrow project, edited by the established scholar Johannes de Laet, printed and published by the renowned houses of Hackius and Elzevier, and of course a direct result of the expedition that was initiated by Johan Maurits of Nassau-Siegen.

A general overview of the results of our census shows straight away that there are relatively many surviving copies of the HNB. So far, we have identified 305 copies of the original edition from 1648, and there is little doubt that more copies are to be found in the stacks of small provincial libraries, museum libraries, and private collections across the world. There is no undisputable evidence on the extent of the original print run. Hoftijzer showed that the printing office of Hackius was capable of producing up to 2100 copies for three volume octavo editions, but without additional information on quantities of paper or production costs for the HNB it is not possible to project these numbers directly on this treatise.[11]

The setup of our census involved three basic steps: locating as many copies as possible, getting them confirmed either through autopsy or with the help of local experts, and gathering information concerning the coloring, binding, and provenance of the copies. Since the HNB was printed in the Netherlands, we started our search for copies simply by checking the Short-Title Catalogue, Netherlands (STCN), which is the Dutch retrospective bibliography for books published before 1801. Then, we listed all copies we could find through search shells such as Worldcat and the Karlsruher

Virtueller Katalog, followed by the copies we came across in national union catalogues, and, occasionally, in local library catalogues.[12] This amounted to a list of approximately 150 copies and a number of epistemological and practical problems connected to the list.

Most importantly, we needed to figure out how we could be sure that these copies on our list actually existed. In the case of the copies that are derived from the STCN, we know that a trained bibliographer described them on the basis of autopsy. That is however not the case for copies found in virtual search engines, such as Worldcat and the Karlsruher Virtueller Katalog, and as a result of data entry from old card catalogues, even local library catalogues have their surprises. It turned out that there are several "ghost copies" of the HNB listed in databases, that is, copies that are accounted for as paper copies but only exist as access to a digitized copy. In other cases, copies were lost or damaged to such an extent that they are no longer available for consultation. The most telling example is a copy in the Herzogin Anna Amalia Bibliothek in Weimar, which was severely damaged in the fire of 2004. The copy was replaced with another copy, bought at an auction in 2007, but this does not erase the fact that the blackened remains of the earlier copy, with specific material evidence connected to it, are in a depot somewhere. Naturally, we listed both the original and the replacement copy in the census. A completely different problem are the copies that are not listed in any online catalogue but undoubtedly exist somewhere in the stacks of a library, museum, or in the collection of a private collector. How do you find books that are not accounted for in any catalogue, bibliography, or inventory? There is, in fact, no better answer than to look systematically and make use of your network of scholars and librarians in different countries.

We sent out a survey to all libraries of which we expected, either on the basis of existing catalogue records or the profile of their collection, that they might own a copy. We examined as many copies as possible ourselves, foremost the copies in the Netherlands, Belgium, and Sweden, with the help of close colleagues who traveled around in Great Britain and Brazil, and through photos and descriptions of local experts and librarians in the other countries. This eventually led to a detailed list of approximately 250 confirmed copies. A final call through the mailing lists of the Consortium of European Research Libraries, the European Botanical and Horticultural Libraries Group, and the Council on Botanical and Horticultural Libraries in June 2018, as well as the decision to include six copies that were, at the time, in possession of antiquarian booksellers, brought the total to more than 300.

Our census was not just intended to locate and identify the copies but also to record the material evidence connected to the copies. Two issues that remained throughout the census were the level of detail that we would apply to the entries and the uniformity of the terminology that we use. When it came to provenance data, we decided not to leave out any information. Every initial, removed bookplate, or illegible name that we know of has been accounted for in the census. We did, however, restrict ourselves when describing the content of annotations throughout the text. One of our initial questions was whether marginal annotations included Linnaean or other forms of taxonomy. Several respondents replied that they were not qualified to answer that question, and since we had no option to check every tome ourselves, we had to drop that from the survey. Uniformity of terminology was particularly problematic when describing the bindings. Some terms, such as parchment and vellum, are used indifferently in catalogues even though they are not the same. The same applies to the difference between gold tooling and stamping and the use of ornaments, fillets, rolls, and so on. Furthermore, descriptions in other languages, such as *pasta española* and *veau blanc*, lose their connotation when replaced by English equivalents. We decided to rely on local expertise where possible and use preferred terms from the Language of Bindings thesaurus in other cases.

Manufacture

One significant outcome of the census is that there are very few printing errors and textual divergences throughout the tome. The fingerprint of the STCN shows that there are no discrepancies at edition level. All known copies have the same collation and the same typesetting. Moreover, there is no evidence of major editorial corrections at the press or typical print-shop accidents that may have required resetting the type for one sentence, paragraph, or an entire page. Even if we zoom in to individual words and characters, we could not find any divergences other than the occasional poorly inked punctuation mark or misprinted page number. In fact, after we had spent days comparing digital and physical copies, we concluded that most of the anomalies we had found were the result of poor digitization standards rather than the printing itself.

   The uniformity of the text demonstrates that the tome was produced with great care, devotion, and skill in a single print run. It underlines that the publication of the HNB was a prestigious project for which only experts and the best craftsmen were hired. The tome was printed on the presses of François Hackius (1605?–1669) in Leiden. Hackius was a typical academic printer and bookseller from Leiden who almost exclusively produced Latin language editions.[13] He had an international clientele and was a regular attendee at the Frankfurt book fair from 1640 onwards. According to Elzevier scholar Alphonse Willems, "the Hackius" – he refers to the family of printers as a whole – were the only Dutch printers who could withstand the comparison with the Elzeviers.[14] In fact, it is likely that François Hackius was trained in the Elzevier company. Relations between the two printing houses were strong ever since, and this is reflected in the publications that were co-produced by both firms. The STCN lists 18 editions that have both the names of Hackius and the Elzeviers in the imprint. There is little doubt that Hackius printed other works for the Elzeviers that were not accounted for in the imprint.

If there ever was any doubt, typographical features prove that the HNB was indeed printed in Leiden by François Hackius. The work contains four head pieces and one tail piece. One head piece that is found in the HNB, depicting a crowned lion surrounded by other animals and a floral pattern, was used in most other folio publications by Hackius after 1648, such as *Monumentum Holmiæ positum Renato Des Cartes* of 1650, Jean Mercier's *Commentarii, in Iobum, et Salomonis Proverbia, Ecclesiasten, Canticum Canticorum* of 1651, and Johannes Calvin's *Institvtionvm Christianæ Religionis Libri Quatuor* of 1654. It is found neither in any publication before 1648, nor in publications of other printers, so it may well have been designed for and only used by Hackius. Of the three other head pieces used in the HNB, one is found predominantly in Elzevier publications from the 1630s to the 1650s, as well as in the only folio edition published by Hackius before 1648, namely Antonius Walaeus' *Opera Omnia* of 1643. The other two head pieces have so far been traced in only one other publication, not surprisingly, printed by Elzevier. The only tail piece found in the HNB is used in numerous publications of the Leiden Elzeviers, probably first appearing in the 1625 edition of *Nieuvve Wereldt ofte Beschrijvinghe van West-Indien* by Johannes de Laet (Figure 3.1).

Wood blocks and type material were of course constantly exchanged between printers. Claims based on typographical evidence should always be made with some reservations, but it is a safe assumption that Hackius had at least three blocks that were originally used by the Leiden Elzeviers in his possession for some time. The exchange of wood blocks confirms that close connections between Hackius and the Elzeviers had existed at least since 1643. The cooperation for the production of HNB was thus completely in line with standing business relations between both printing houses.

When we look at the other illustrations in the tome, it is striking that the dominant technique is woodcut. The well-known engraved title page, depicting Tapuya Indians in a Brazilian forest, is a copper engraving, but the illustrations of plants, animals, and the peoples of northeastern Brazil throughout the work are all woodcuts. The HNB was published at the time of a technical transition in the use of illustrations. In the first half of the seventeenth century, most publications in the field of natural history were illustrated with woodcuts. This changed in the second half of that century, when engraving became the dominant technique. The transition in technique is reflected in the production and use of illustrations in the HNB. The cultural-historical background of this transition falls outside the scope of this chapter; however, it is important to keep in mind that while engraving provided more detailed illustrations, the cost of an engraving in the first half of the seventeenth century was up to ten times that of a woodcut illustration.[15] There is no indication that the use of engravings was considered at a certain point. Even for a highbrow project as the HNB, the production costs were probably too high.

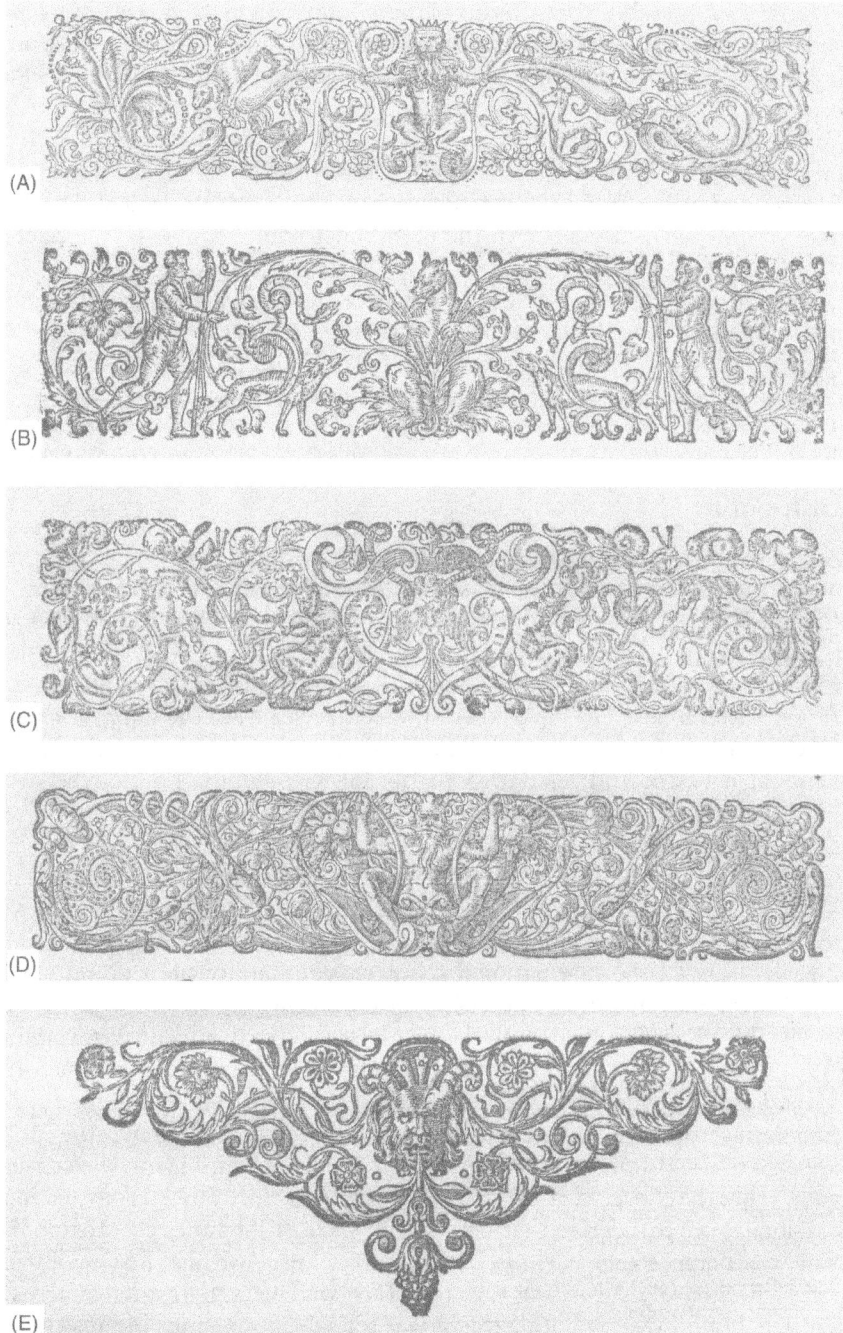

Figure 3.1 Four head pieces (A–D) and one tail piece (E) used throughout the *Historia Naturalis Brasiliae* (Piso and Marcgraf, 1648). Leiden University Libraries (copy 1407 B 3).

Little is known about the afterlife of the woodblocks in the seventeenth century, but at least some of them have survived much longer. The Rijksmuseum holds four catchpenny prints from the late eighteenth and early nineteenth centuries, which were printed from the original woodblocks from 1648.[16] One print from late eighteenth century shows the Indigenous people depicted on page 248 in the Marcgraf part of the HNB, with the imprint Johannes Bouwer and David le Jolle. The same print was reissued by Clement, De Vri, and Van Stegeren in Zwolle in the early nineteenth century. Other catchpenny prints make use of different woodblocks, showing the original illustrations in new contexts with added paratexts. So far, we have identified just one animal: the woodblock of the llama on page 243 of the HNB was reused on a print that further holds animals from Conrad Gessner's sixteenth-century encyclopedic work *Historiae Animalium*.

Distribution

Apart from the business relations between Hackius and Elzevier, there might have been a practical reason to print the HNB in Leiden. Editor Johannes de Laet was based in the city so he could monitor the process at the workshop of Hackius from close range. De Laet had worked with the Elzeviers since 1625 and was the editor of several volumes of the celebrated Elzevier *Republics*. The inclusion of Louis Elzevier III in the imprint, arguably as the publisher of the book, was undoubtedly motivated by the international network and reputation of the Elzevier family. Louis, who just like François Hackius started his own business in 1638, had traveled to Denmark in 1632, 1634, and 1637, and to Italy in 1636. This had helped him to build up relations with scholars such as Johannes Meursius, Isaac Vossius, Ole Worm, Jean and Arnold Corvinus, and Lucas Holstenius, next to the vast network of the family business that he could already rely on. The Elzeviers served the international scholarly community on a grand scale and had their agents in all corners of the continent.

The HNB is first listed in the retail stock catalogue of Elzevier in Amsterdam in 1649. In the same year, the treatise is also mentioned in a catalogue of Melchior Martzan, who was the caretaker of Johannes Janssonius' office in Copenhagen. Somewhat later, in 1659, the title appears in the catalogue of the Frankfurt office of Joan Blaeu. The census shows that the Moretus family had a copy at least since 1675. From auction catalogues we can derive that booksellers such as Johannes (I) Janssonius van Waesberge, Pieter Niellius, and Johannes Verhoeve also had a copy in stock. It is hardly surprising that the HNB was instantly available across Europe, but it is worth noting that it was for sale in the shops of Janssonius and Blaeu, two of the traditional Amsterdam competitors of Elzevier.

When we take a closer look at the early stages of the distribution of the HNB, four groups of early possessors of a copy stand out in the census:

the libraries of religious orders, such as abbeys and Jesuit colleges, medical doctors and naturalists, European royals and aristocrats, and finally collectors from a civic elite. The census includes seventeenth-century provenances for Jesuit orders in Ghent, Vienna, Cologne, Lyon, Münster, and possibly Naples, the Barefoot Carmelites in Barcelona, and the abbeys of Ninove, Grimbergen, Afflighem, and Elwangen. The provenances for most of these copies date from the first 20 years after the treatise was published. One of the copies that is now in the National Library of Belgium was, according to a manuscript annotation, already in the possession of the Jesuit College in Ghent in 1648. The copy now at Jesus College Library in Oxford was bequeathed by Edward Herbert of Cherbury, who died in the year the HNB was published. The relation between the Elzeviers and Jesuits has been recognized before, while the practice of medical science at Jesuit schools and in monasteries and abbeys is of course well-known.[17]

The second group of early owners includes the English physicians John Goodyer and Christopher Merret, French medical doctor Louis Morin de Saint-Victor, the Italian court physician Romolo Spezioli, and the Germans Emanuel Brigel, Lukas Schröck, and Martinus Fogelius. From correspondence with Elzevier, we know that Danish naturalist and collector Ole Worm anxiously waited for the arrival of his copy in 1649, while auction catalogues reveal that Anchises Andla, court physician to William Frederick, Prince of Nassau-Dietz, Leiden professors of medicine Johannes Walaeus and Adolphus Vorstius, Harderwijk professor of medicine Franciscus Jacobus Cochius, doctors of medicine Samuel Coster, Gerard Calf, Wolfius Schonevelt, and Johannes de Vogelaer all had a copy in their possession. The data that we have taken from auction catalogues (Brill Book Sales Catalogues Online) shows that this information is complementary to the material evidence that we recorded from the tomes. It also poses the question: where are these copies now? None of these names that we derived from auction catalogues are listed in the census. This does not necessarily mean that these copies have been lost. It is not unlikely that several early possessors left no traces in their copies. They may well have survived in a collection without any material evidence of their initial owners.

Furthermore, we can identify a number of copies that were part of the collections of the European aristocracy. Gaston d'Orléans owned a copy, as did Cardinal Jules Raymond Mazarine, Prince Eugene of Savoy, and Johann Friedrich von Braunschweig-Lüneburg. There is a copy with the supralibros of King James II, and one with the wax seal of Swedish commander Magnus Gabriel De la Gardie. The colored copy in the Royal Library in Copenhagen was probably donated by Johan Maurits to King Frederick III of Denmark.[18] There is no evidence that Johan Maurits sent other copies to European royals as a gift, but it is certainly imaginable that he did.

In the Dutch Republic, stadtholder William Frederick, Prince of Nassau-Dietz, gave a copy to Franeker University in 1649, with an additional folium containing a printed dedication (Figure 3.2), while the States of Groningen

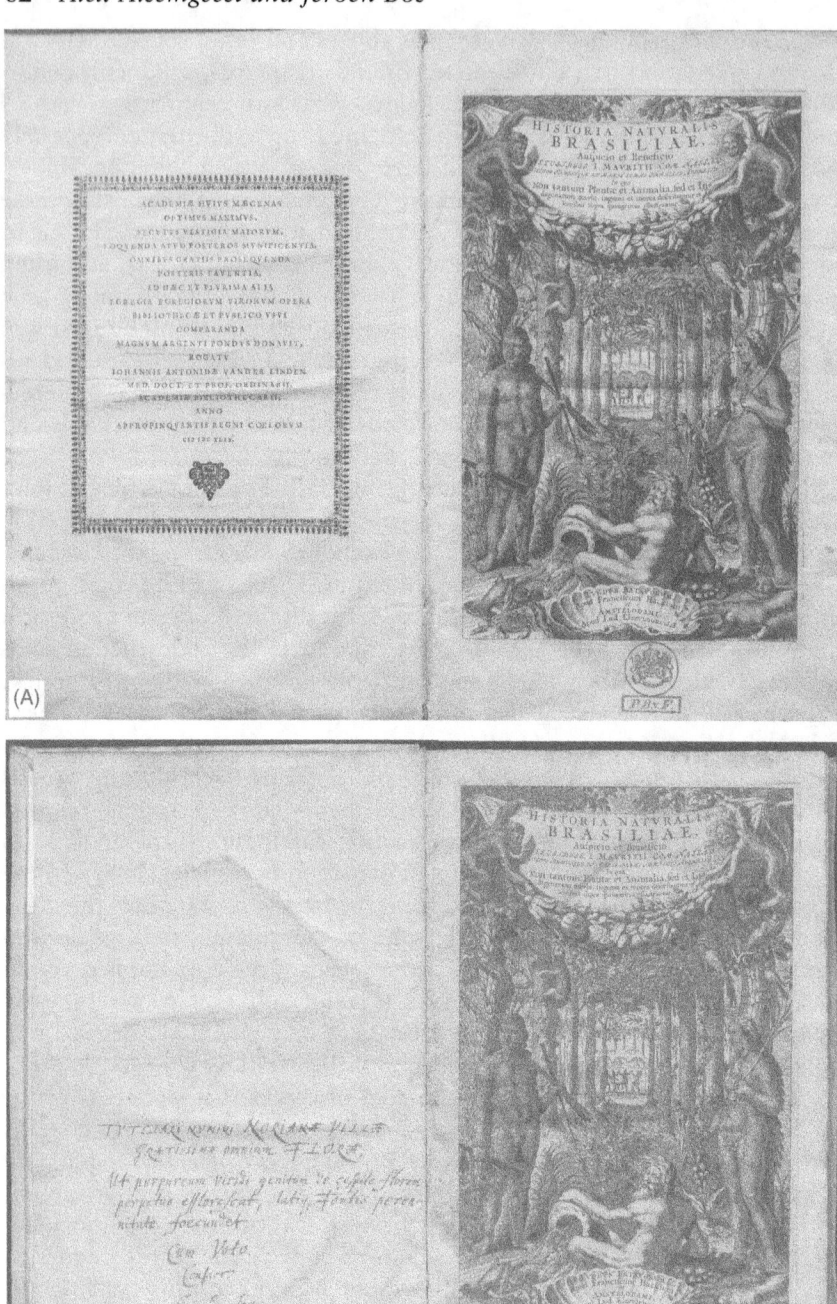

Figure 3.2 (A) Printed dedication of William Frederick of Nassau-Dietz to Franeker University Library. Tresoar (copy 700 Ntk fol). (B) Unidentified manuscript annotation. Uppsala University Library (copy Leufsta Collections F 92).

donated one to the University of Groningen in 1668. Both donations underline the prestige connected to the treatise. Other copies can be connected to the collections of the Fagel family, Grand Pensionary of Holland Adriaan Pauw, and collector Johannes Thysius. Furthermore, auction catalogues show a range of other possessors from various backgrounds, such as Leiden minister Jacobus Derramoud, commander of Purmerend castle Jacobus Franciscus Hinlopen, Leiden minister and president of the Theological College of the States of Holland of Leiden University Martinus Ubbenius, Leiden jurist and town clerk of Breda Janus Vlitius, Leiden magistrate Arnold Wittens, and Leiden professor of oriental languages and mathematics Jacob Gollius.

The diversity of people who owned a copy of the HNB in the first decades after its publication, ranging from a civic elite, medical professionals, book collectors, universities, Jesuit schools, and kings and princes, suggests that the scope of the publication was broad. It was a treatise for the natural and medical sciences, for geography and a more general interest in travel and exploration, but certainly also for collecting and prestige. This is furthermore reflected in the contemporary bindings that have survived. The dominant binding seems to have been Dutch parchment, which was typical for the time, with the ratio between parchment and leather being approximately 70 to 30. Many copies have some form of gold tooling or blind stamping. Several copies that were bound in leather have an elaborate panel design. A small minority of copies has been personalized with a coat of arms. These are often the copies with a royal or aristocratic provenance.

Coloring

Next to the binding and decoration, coloring of the images was one of the most noticeable adaptations that an owner could make to his copy. Coloring was, however, expensive and time-consuming and not necessarily appreciated as it is today. There are in fact very few colored copies of the HNB and little is known about the process. A specific question in the survey was whether the illustrations in the copy held by the institution were colored or not. Before the conducted census, six colored copies were known. In their pioneering work *A Portrait of Dutch 17th Century Brazil*, Peter Whitehead and Marinus Boeseman identified colored copies in the Library of Congress (Washington), the collection of the late Joaquim de Sousa-Leão (Rio de Janeiro), Universiteitsbibliotheek (Leiden), Rijksmuseum van Natuurlijke Historie (Leiden), Kongelige Bibliotek (Copenhagen), and the Staatsbibliothek (West Berlin).[19] The names of some of these institutions were altered over the years, but not much changed concerning the location of the copies. The Rio de Janeiro copy is the exception: it was relocated to São Paulo when the collection of Joaquim de Souza-Leão (1897–1976), a diplomat and Brazilian ambassador in the Netherlands in the 1950s, was acquired by Banco Itaú in 2002 and transferred to the Instituto Itaú Cultural.

Following the census, we can now name eight more copies with colored illustrations, namely in the Universiteitsbibliotheek (Ghent), Bibliothèque Nationale (Paris), Landesbibliothek (Oldenburg), Trinity College Library (Dublin), British Library (London), Royal College of Physicians (London), Public Library (New York), and Missouri Botanical Garden Library (St. Louis). Furthermore, there are two copies of which only the engraved title page is colored: one at the McGill University Library (Montréal) and the other at the Natural History Museum (London). This brings the number of colored copies to 14 on a total of just over 300 copies. What statements can be made and what questions should be asked based on these findings? Can we say anything in particular about where, when, and how the coloring was done? Autopsy and detailed analysis of all colored copies falls outside the scope of this census, but there are certainly some remarks to be made about the findings.

First, we will discuss the colorists of the tomes, in Dutch called "afsetters." According to art historian Truusje Goedings, the study of the profession, production, and identification of individual colorists is still largely unexplored territory, owing to the fact that early modern book coloring has been looked down upon for a very long time.[20] From the middle of the eighteenth century, the coloring of book illustrations had a more industrial character and more is known about the relationship between publishers, printers, and colorists. For the sixteenth and seventeenth centuries, information and sources are scarce. The profession of colorist was not bound to rules of an organization, such as a guild. Anyone could express himself as colorist and it was one of the few career opportunities open to women. Colorists were considered artisans, not artists, and most carried out their work anonymously. Only a few gained such reputation that their names are known, some even were considered "meester afsetter," or master colorist. It is, however, unclear whether they had to pass a test to receive this title, which would suggest some form of organization, or that it was given as token of respect for excellent craftsmanship.

One such master colorist was Amsterdam-based Frans Koerten (1599/1600–1668). When he died, his stock of books and prints was auctioned. The printed auction catalog reveals that Koerten possessed a colored and uncolored copy of the HNB 20 years after the treatise was published.[21] This suggests that Koerten was the colorist for at least some of the copies, and that the coloring took place over a period of several decades. Several books in the same catalogue are partially colored, indicating that Koerten was working on them. The entry for the HNB in the catalogue holds the remark that the colored copy was "curiously colored after the princely original" ("volgens 't Prinselijck Originael, curieus afgeset"). It is uncertain though what the original copy is and to whom the title of "prince" refers. It is very tempting to designate a master copy of the HNB which functioned as example for other copies. It is likewise very tempting to consider Johan Maurits as the prince whose copy was the referred to as the "princely original," but is there a historical basis for this consideration? In 1652, the count was ennobled and was since entitled to be addressed as prince.[22] In that case, the colored copies in Berlin and Copenhagen have a provenance that goes back to Johan Maurits and would be possible candidates. It is very questionable though whether the auctioneers followed this line of thought. Auction catalogues were often compiled in haste, and the auctioneers seized every opportunity to praise the books and prints for potential buyers.

Even though we have not examined all 14 colored copies in detail, it is clear that there are notable differences between them. Already on the title pages, it is striking that different colors have been used for the birds in the right top corner. On the other hand, there are some striking similarities in all copies that we have seen. The vase at the bottom of the title page, for instance, appears to be red and yellow in every copy. Moreover, the animals coming from the vase show a high level of resemblance in every examined copy. This cannot be coincidental. More research is needed about the detail and quality of the coloring of individual copies, arguably in combination with the *Libri Picturati* that hold the original drawings by Eckhout, before anything conclusive can be said about the coloring. However, if we focus on the similarities rather than the differences, there is reason to assume that the majority of copies were colored after a master copy, and given the quality of the coloring in the copies that we have seen, presumably by Frans Koerten in Amsterdam.

Minor variations, especially on the title page, might be explained by preferences of the patron or even the colorist. More comprehensive variations throughout the tome might be an indication that the coloring was not done after a master copy, probably not even by Frans Koerten. Not all colored copies were necessarily handled by professional colorists. It was a fully accepted pastime for early modern book owners to color the illustrations themselves. Instructional coloring manuals which contained recipes for preparing pigments and watercolor were highly popular.[23] The copies at the McGill University Library and the London Natural History Museum, with only the engraved title page in color, could very well be the work of enthusiast book owners. It might also be a stub from colorists who owned a copy and did not want to risk coloring the whole work without a guaranteed resale. By coloring the engraved title page, they could tempt interested buyers to buy a fully colored copy.

One of the well-known features of early modern coloring of book illustrations is color bleeding. There are several reasons why colors bleed. It was advised to not directly apply the colors on the paper but pre-process it with a glue-like substance. Another, rather obvious, reason for bleeding is the quality of the paints used. Lesser talented and beginning colorists did not have the financial means to invest in high-quality pigments. Paint that was applied too quickly tended to clump together. This is exactly what happened with a colored copy from the collections of the Library of Congress. In the process of opening the pages, the illustrations were damaged.

Finally, can we say anything conclusive about the early owners of the colored copies? Two of the colored copies have a provenance that may be traced back to Johan Maurits. The colored copy that is now in the Staatsbibliothek zu Berlin might have been his own, while the copy in the Royal Danish Library was presumably a gift by him to Frederick III. Most other colored copies lack a clear provenance in the seventeenth century. The copy at Trinity College Library that belonged to the Dutch Fagel family might qualify, though it is uncertain when the Fagels acquired it. The copy at the Royal College of Physicians came from the English collector Henry Pierrepont in 1680. The trajectory of most other colored copies only becomes apparent once they are in the hands of bibliophiles, such as Gilles-Jean Rooman, Georg Friedrich Brandes, James Carson Brevoort, and Ferdinand Casper Koch. Also, the copy in the British Library, owned by Hans Sloane, shows no traces of use and seems to have been a copy for presentation rather than for study.

Looking at the coloring from a different perspective, that is, by looking at the uncolored copies rather than the colored ones, we can state that copies that evidently left the Dutch Republic right after publication are all uncolored. This supports the idea that copies were colored in Amsterdam in the first two decades after publication. The same arguably applies to the copies in royal and aristocratic collections throughout Europe, which were generally uncolored. Surely it can be interpreted as a conscious decision by the custodians and librarians of these collections to acquire an uncolored copy, but it would have been hard to color these copies even if they had wanted to. Atlases and emblem books could be colored according to preferences. But how were you supposed to color all these exotic plants and animals "*ad vivum*" without original drawings, a master copy, or any indication what their natural colors were?

Redistribution

How soon the process of redistribution began is obvious when we bring to mind that Edward Herbert of Cherbury bequeathed a copy in 1648, the year the HNB was published, and that professor of medicine Johannes Walaeus' copy was auctioned in the following year. The copy of Edward Herbert of Cherbury was bequeathed to Jesus College Library in Oxford, where it is still today. Other copies, however, transferred ownership multiple times. The census brings forth hundreds of these second, third, or later generations of owners. We will give two examples from the eighteenth century: the naturalists and the bibliophiles.

Since the HNB remained the principal scientific text on the natural history of Brazil throughout the eighteenth century, it seems only logical that several notable naturalists from that era were among the new owners. Carl Linnaeus was one of them. He included some descriptions by Piso and many more by Marcgraf in his *Systema Naturae*. Linnaeus' own copy is now in the library of the Linnean Society of London and has generic identification of some botanical figures in Marcgraf's section. The copy of his good friend and court physician Abraham Bäck is now in the Hagströmerbibliotek in Stockholm. The same library also holds a letter from Daniel Rolander, dated 20 May 1756, where he thanks Bäck for the loan of "Piso" and some other books on his trip to Suriname. It is a fascinating thought that Linnaean apostle Rolander had a copy of the HNB in his baggage when he was collecting plants and insects in roughly the same geographical area. The census shows that there were many more in Swedish naturalist circles that owned a copy, such as entomologist Charles De Geer and the Bergius brothers.

Naturalist interest in the HNB was of course not restricted to Sweden. In England there are copies that bear the names of James Petiver, Hans Sloane, and Joseph Banks, in France there is a copy of Georges Cuvier, in Italy there are copies connected to the botanical gardens of Padua and Bologna, as is the case in Edinburgh, Amsterdam, and Brussels. In general, the copies with a clear naturalist provenance show traces of intensive use, even if it is not always clear who was responsible for them. The copy that is now at Universidade Estadual de Campinas (Unicamp) has a bookplate from the Reformed School in Frederica, Denmark. The copy is annotated throughout and has five inserted illustrations. One of them is instantly recognizable as the lemming from Ole Worm's *Museum Wormianum*. At first glance, it seems out-of-place to include the image of a Nordic animal in a book on Brazilian nature, even though it is inserted in a section with small quadruplets. The Danish provenance makes it comprehensible why and at what point the treatise was studied in relation to the works of Danish naturalist Worm.

The census further shows a rise of bibliophile interest in the HNB in the eighteenth century. The concept "bibliophile" is ambiguous, since the word was not used in English until the nineteenth century, but collectors have been around as long as there have been books. Nonetheless, the increasing interest in antiquarian books for other purposes than the study of their content, in combination with the availability of great quantities of old books due to the dissolution of monastic libraries, paved the way for the grand collectors of the late eighteenth and early nineteenth centuries. A textbook example of this history is the copy of the legendary Belgian bibliophile Charles van Hulthem, that came from the Abbey of St. Cornelius and St. Cyprianus in Ninove.

Heritage, Patrimony, and Survival

The census is instrumental in demonstrating the life cycle of the typical copy of the HNB. After private ownership in the seventeenth and eighteenth centuries, a large part of the copies was acquired by societal or academic libraries in the nineteenth and twentieth centuries. This process of mass transition of book ownership, from the private to the (semi) public sphere, has only recently been given proper attention. Book historian Pierre

Delsaerdt uses the French term "la patrimonialisation" to describe the process of transition where book collections were taken from their original point of collection and incorporated into larger, mostly national, inventories, such as National State Libraries or University Libraries.[24] The term is not new, Delsaerdt argues, but mainly used by French cultural historians in museum studies to describe the more known transitions in art and architecture. The mass transitions with books, which took place in the same period, have long been overlooked but show very similar patterns. Likewise, books became heritage objects, losing their original function as reference works or objects of study. In the case studied by Delsaerdt, which concerns the acquisition of the library of the abbey of Tongerlo by the government of the United Kingdom of the Netherlands in 1827 for the sum of 8,000 guilders, the transition was "friendly," although some persuasion was needed to convince the Tongerlo clergy.

More often, though, these mass transitions of book collections had a hostile character, especially in times of crises, such as the French Revolution or the successive Napoleonic Wars, which not only shook Europe politically, militarily, and economically but also culturally. In their attempts to open royal, private, and monastic libraries for the benefit of the nation, the French revolutionaries could not foresee the devastation of previous book networks and inventories, carefully structured over time. The same is true for the so-called "artistic conquests" against defeated, occupied, and annexed countries in the years following the Revolution.[25] According to Dominique Varry, millions of books were dispersed, raided, or otherwise jumbled up in the 14 years from 1789 until 1803. In even less time, from 1782 until 1787, the ecclesiastical reforms under Austrian emperor Joseph II had a devastating effect for the monastic libraries in the Habsburg lands.[26] The closures of monasteries led to the confiscation of their libraries. Although the Court Library (Hofbibliothek) in Vienna had the right to select the most precious books of the closed monasteries, the staff could not handle the mass transitions, leading to the auctioning or downright destruction of centuries old collections. Medieval manuscripts and incunables were sold to paper buyers who had no eye for their contents. These kinds of hostile confiscations took place all over Europe and with more or less the same argument: the invaluable collections should no longer be in the exclusive hands of the clerical, economic, or political elite but accessible to anyone.

Patrimonization was not just a process of confiscation. Several important private book collections that had been carefully built up in the late eighteenth and early nineteenth centuries were either donated or sold to national or university libraries. In the census, we find examples of this process throughout the nineteenth and twentieth centuries. Some copies have a rare combined history of confiscation and legal acquisition. The copy in the National Library of Belgium that belonged to the Jesuit College in Ghent was first confiscated by the French in 1795, transferred to the Ecole Centrale du Département de la Dyle (Central School of the Dyle

Department) in 1797, and subsequently donated to the City of Brussels in 1803. That library was acquired by the Belgian State for the Royal Library of Belgium in 1843.

The transfer of copies from private to (semi) public ownership meant that the use of the tome altered. Copies that were acquired by libraries and museums became heritage objects and were no longer in circulation. One would expect that this is the end of the material history of the tome. However, typically a library stamp was added to the tome, sometimes the tome was restored, or a new binding was made. Moreover, the decision in most libraries to place the tome in a special collections department, to take it out of regular circulation, to put it on display, or to lend it to a museum are all examples of heritagization. Some copies, moreover, defy the typical life cycle by returning to private ownership. The fact that the copy of the Muséum National d'Histoire Naturelle with annotations by Georges Cuvier is now for sale at an antiquarian bookseller is illustrative. Other copies transferred from one institution to the other as a result of the merge, dissolution, or renaming of libraries. We will highlight some copies of the HNB to illustrate the dynamics of institutional life.

The Copy of Boudewijn Büch at Teylers Museum

The Teylers Museum is an art, natural history, and science museum, located in the Dutch city of Haarlem. It was originally founded in 1778 after the wealthy cloth merchant Pieter Teyler van der Hulst (1702–1778) donated his fortune for the advancement of contemporary art and scientific studies. The acquisition and storage of a book collection was thought necessary for reference and inspiration. The Teylers Museum holds two copies of the HNB, one in a contemporary binding, the other in a later, possibly eighteenth century, green morocco binding. The first copy was probably acquired very early in the museum's existence. No provenance can be ascribed to this copy, but the nineteenth-century library stamp shows that it has been in the collection for quite some time. The other copy portrays a long list of interesting owners and seems to defy the general trajectory of most HNB copies.

The last owner was Boudewijn Büch (1948–2002), a renowned Dutch novelist and television presenter who collected many objects of natural history, ranging from naturalia to rare books. At the time of his death, his library was estimated to hold 100,000 books. The auction catalogue, effectuated by auction house Bubb Kuyper, consisted of three volumes. Before his library, known as Bibliotheca Didina et Pinguina, went to auction, Teylers Museum was able to acquire some volumes which the museum did not already possess. Among the volumes that were picked by former curator Bert Sliggers was a copy of the HNB. Due to time pressure, Sliggers was unaware of the fact that the Teylers Museum already possessed a copy of the HNB and thus selected the Büch copy for the museum.[27]

The copy includes three book plates of previous owners, the first being that of the Polish poet and chamberlain to the Polish court Thomas Catejan Wegierski (1755–1787). Because of his satire and licentious lifestyle, he had to leave the court and traveled to Italy and France before arriving in Philadelphia. It is not known what happened to the tome after the premature death of Wegierski, but the booksellers' ticket of Rey et Gravier suggests that it must have been sold by these Paris book dealers between 1815 and 1839. A manuscript annotation states that it was in the possession of American physician and professor of medicine Walter Channing (1786–1876) on 11 April 1840. He sent it as gift to Amos Binney (1803–1847) and included a two-page handwritten letter to accompany the donation. Binney was an American physician and malacologist and co-founder of the Boston Society of Natural History in 1830. The Society replaced the Linnaean Society of New England which only existed between 1814 and 1822. After his death, the widow of Amos Binney bestowed his book collection to the Boston Society of Natural History.

With this transaction, the tome went from private to institutional hands. Normally, this would mark the final stage in the circulation of the tome. But this HNB copy remained in motion, because of declining public interest in the outmoded presentation of natural history collections and, subsequently, the poor financial situation of the Society in the 1940s. The Boston Society of Natural History changed its name to the Boston Museum of Science in 1951 and shifted its focus completely from propagating and facilitating scientific research to popular education. Five years before, in 1946, the extensive library collection was sold.[28] This is how this particular copy of the HNB recurred on the market. It is unknown who owned the copy between 1946 and 1993. In that year, the Nürnberg-based German art and bookseller Kistner offered the HNB for the sum of 12,000 DM. It was Boudewijn Büch who eventually purchased the tome. He would loan it for an exhibition about his extensive collection at the Natuurhistorisch Museum Rotterdam in 2001, less than a year before his untimely death.

Three Copies at Naturalis Biodiversity Center

The library of Naturalis Biodiversity Center (Leiden, The Netherlands) holds three copies of the HNB, one of them being a colored copy. Naturalis was founded in 1820 by royal decree under the name 's Rijksmuseum van Natuurlijke Historie (National Museum of Natural History) and one would expect that the HNB copies were part of the collections from the beginning. This was not the case: the first copy was only acquired in 1975. The acquisition was research-driven, fulfilling the explicit wish of the museum staff to own a copy for reference, after a large-scale investigation into the zoological taxonomy of the depicted and described "Brazilian" animals in the HNB was carried out.

The story of the iconography of Dutch Brazil and its complex examination by many scientists over the years has been extensively written about elsewhere. In short, and as far as it concerns the staff of Naturalis, the iconographical search started when professor Enrico Schaeffer (1907–1979), an art historian from Rio de Janeiro, contacted the Leiden museum to help him with the identification of species in the HNB. Schaeffer was the organizer of an exhibition about the visual legacy of Johan Maurits' governorship. This exhibition took place in 1968 and truly exceptional loans of animal drawings from the Russian Academy of Sciences were displayed. As an art historian, he felt insecure in identifying the animals on the drawings and by 1972 he decided to contact the Leiden staff.

This call for assistance sparked a prolonged iconographical and taxonomical study, led especially by ichthyology curator Marinus Boeseman (1916–2006). Over the course of several decades, this would lead to many publications.[29] Boeseman would find a companion in Peter Whitehead from the natural history department of the British Museum. Together, they are responsible for pioneering work in the field of the iconography of Dutch Brazil, combining their thorough investigations to find traces in archives and libraries that, considering Cold War policies, were difficult to access, like the Russian Academy of Sciences in St. Petersburg and the Jagiellonian Library in Kraków, Poland. In 1977, this library confirmed strong assumptions that it was in possession of the *Libri Picturati*, a large set of botanical and zoological drawings which was presumed missing since the Second World War.

Boeseman and Whitehead cooperated with many colleagues in the field of natural history, one of them being the Leiden curator of crustacea Lipke Bijdeley Holthuis (1921–2008). Boeseman could make extensive use of Holthuis' library for reference. In an article on the overlooked information about Brazilian zoology in Caspar Barlaeus' *Rerum per Octennium in Brasilia et alibi nuper Gestarum sub Praefectura* (1647), Boeseman clearly acknowledged his indebtedness. He stated that "it would have been impossible to achieve the present result" if he had not been able to consult the "valuable items in the extremely rich library of my colleague Dr. L.B. Holthuis, emeritus curator of Crustacea in the Leiden museum, kindly put at my disposal."[30] This "extremely rich library" comprised some 8,000 book volumes. As a devoted collector, Holthuis had very good relations with antiquarian booksellers worldwide.

In 1974, Boeseman mentioned the possibility of acquiring a colored copy of the HNB to the museum's management. To possess a copy of the tome was considered a welcome asset for the ongoing iconographical and taxonomical research conducted on Dutch Brazil. In order to hastily obtain the financial means for this purchase, the museum decided to sell a duplicate set of P. Bleeker's *Atlas Ichtyologique des Indes Orientales Néerlandaises* (1862–1878). The copy of the HNB that the museum acquired came from the collection of the Rotterdam politician, amateur historian, and book collector Ferdinand Casper Koch (1873–1957). After his death, his collection was sold by the German auction house Hauswedell & Nolte. In 1959, a first, rather small, auction was organized, to be followed by a proper auction of the entire collection in December 1974.

The colored copy has been extensively used for research ever since its acquisition, but the tome also served another purpose, namely as heritage, cultural memory, and consequently to facilitate the public outreach of the museum. As one of the few colored copies, it is an interesting object to put on display. In 2014, the tome was loaned to the Mauritshuis in The Hague as part of the reopening exhibition, following a long renovation period, of the fine arts museum. The tome was part of a section of the exhibition that recounted the first owner of the museum building: Johan Maurits.

Naturalis Biodiversity Center also possesses two uncolored copies of the HNB that came to the library much more recently. The first one was part of the aforementioned library of Holthuis. He bequeathed not only his book collection but also his complete scientific archive to the museum in 2008.[31] This archive is a rich and largely untapped source of how research in the field of natural history was conducted in the second half of the twentieth century. The Holthuis copy contains an unclear manuscript name dated 1791 and the signature of the Irish physician Sir Thomas Molyneux (1661–1733). In 2014, a third copy was added to the collections of Naturalis when the Netherlands Entomological Society (Nederlandse Entomologische Vereniging) moved its library to Naturalis as a long-term loan. This particular copy contains a signature from the Prussian-born botanist Caspar Georg Carl Reinwardt (1773–1854) and a library stamp of Dutch entomologist Hartog Heys van Lier (1821–1870).

The Brazilian Copies

One of the arguably surprising results of the census is that there are at least 20 copies of the HNB in Brazil today. Most of them were acquired by libraries in the course of the twentieth century or even more recently. It shows that the history connected to the HNB is not just that of Johan Maurits, the West India Company, and seventeenth-century European medicine, but also that of Indigenous people, their knowledge systems, and the history of Brazil. The recognition of Indigenous cultural memory in the text, and arguably the importance of the treatise for present-day Brazil, is omnipresent in the material history of the copies that are now in Brazil. The copy at Campinas, for example, was acquired by Unicamp upon the foundation of the university in the 1960s, at the specific insistence of its first rector magnificus professor Zeferino Vaz (1908–1981). Copies in Brasília and Belo Horizonte were bought by their respective libraries in 1963 and 1979. Half a century before, Brazilian industrialist Julio Benedito Ottoni (1857–1926) donated a copy to the national library in Rio de Janeiro. Brazilian bibliophile and specialist in Tupi-Guarani language Frederico Edelweiss (1892–1976) donated his copy to the Universidade Federal da Bahia in Salvador. Recent donations include the copy of entomologist Johann Becker (1932–2004) to the Biblioteca do Museu Nacional and the copy of Brazilian journalist and bibliophile José Mindlin (1914–2010) to the Universidade de São Paulo.

Only listing the trajectory of these copies implies that the tome has long been considered an important cultural object for Brazil. At first, the importance was arguably more its practical application. As we have seen, Daniel Rolander took a copy with him on his journey to Suriname in the eighteenth century. Nineteenth-century naturalists who operated in Brazil, such as the German Theodor Peckolt (1822–1912), naturally owned a copy. They needed it as reference material. It would be interesting to find out whether this still was the case with Brazilian medical doctor and professor at the School of Medicine of Bahia Egas Moniz Barreto de Aragão (1870–1924), who also owned a copy. More research is needed about the motives of benefactors, but it seems that later donations and acquisitions were not just driven by the practical applications of the treatise, but by an understanding that the tome represents an important part of the cultural memory of Brazil. It is illustrative that the census shows dozens of active acquisitions of the HNB in Brazil, as well as in the United States, over the last decades, and hardly any in Europe.

The current geographical dispersion of HNB copies raises more questions. Looking at the list of all locations where a copy of the HNB is present, one cannot help but notice that, to our knowledge, no copies are present on the African continent. The text has been recognized as holding all sorts of information on language and the natural world, not only of Indigenous people in Brazil but also of enslaved African people. This is a relatively new approach to the treatise that is not yet reflected in the results of the census. We have seen, however, that copies continue to be transferred between collectors and institutions, and it is only logical that a future census will mirror new approaches to the tome.

Conclusion

Sidney Lee's census of Shakespeare's first folio copies was published in 1902, but copies have been added to the list ever since. In 2014, a copy was found at the public library in St. Omer, near Calais; two years later, another one was discovered at a stately home on the Scottish Isle of Bute. The fact that copies of the most sought-after book in the world still turn up more than a century after the search began reminds us that copy censuses are, by nature, imperfect. The 305 copies of the HNB that have been listed in this census probably make up some 20–30 percent of the entire print run. It is highly implausible that the remaining three-quarters of the print run have been lost. Unrecorded copies are bound to turn up in town libraries, at stately homes, or in the vault of a private collector sooner or later.

The copies that we have listed so far can be connected to the different stages of the life cycle of the tome and the multiple histories connected to it. Every copy has a unique history that is partially revealed by the material evidence connected to the copy. Book plates and inscribed names give away some of the former owners, faint numbers on the spine might be shelf marks from an earlier collection, bindings might tell if the tome was a work copy, meant to be annotated, or a luxurious copy that was placed in an aristocratic or bibliophile collection. The stories connected to the material evidence of individual copies are seemingly endless, however, it is not until we list hundreds of these copies that we begin to see patterns. One copy with a seventeenth-century Jesuit provenance is interesting, but if you find seven or eight of them, you start asking broader questions. Conversely, the census can also be used to study what is not listed. Why are there no tomes in a certain library, country, or continent? Or why is the name of a specific naturalist or collector not listed, even though it is hard to believe that they did not own a copy?

In this chapter, we have highlighted some of the basic characteristics that came forth from the census. We are hopeful that other scholars that study the treatise from other perspectives will notice entirely different things. Not only to find new copies, record material evidence, and identify more names, but especially to connect it to other forms of research. One of the interesting options for future study would be to combine the evidence from the census with information from auction catalogues. This might reveal some of the histories that are now nothing but an auction number on the inner boards of a copy somewhere in the special collections of a library. If the census holds one promise, it is that every detail is important in uncovering the story of the book from cover to cover.

Notes

1   Neil Safier, "Beyond Brazilian Nature: The Editorial Itineraries of Marcgraf and *Piso's Historia Naturalis Brasiliae*," in *The Legacy of Dutch Brazil*, ed. Michiel van Groesen (Cambridge, UK: Cambridge University Press, 2014), 173–174.

2   Peter J.P. Whitehead and Marinus Boeseman, *A Portrait of Dutch 17th Century Brazil: Animals, Plants and People by the Artists of Johan Maurits of Nassau* (Amsterdam: North-Holland Publishing, 1989); Marinus Boeseman, Liepke B. Holthuis, Marinus S. Hoogmoed, and Chris Smeenk, "Seventeenth Century Drawings of Brazilian Animals in Leningrad," *Zoologische Verhandelingen* 267 (1990): 1–189.

3   See Alsemgeest and Bos, "Appendix," in this volume.

4   Robert Darnton, "What Is the History of Books?," *Daedalus* (Summer 1982): 65–83 reprinted in Robert Darnton, *The Kiss of Lamourette. Reflections in Cultural History* (New York, NY: Norton, 1990); Thomas R. Adams and Nicolas Barker, "A New Model for the Study of the Book," in *A Potencie of Life: Books in Society: The Clark Lectures 1986–1987*, ed. Nicolas Barker (London, UK: British Library, 1993), 5–43.

5 Pierre Nora, *Les Lieux de Mémoire*. 7 volumes (Paris: Gallimard, 1984–1992); Françoise Choay, *L'Allégorie du Patrimoine* (Paris: Editions du Seuil, 1992); Pierre Delsaert, "De Verzegelde Kisten van de Vrouwe Adriana: De Abdijbibliotheek van Tongerlo en de Patrimonialisering van het Boek in het Verenigd Koninkrijk der Nederlanden," *Jaarboek voor Nederlandse Boekgeschiedenis* 25 (2018): 129–149.

6 David Pearson, "The Importance of the Copy Census as a Methodology in Book History," in *Early Printed Books as Material Objects: Proceedings of the Conference Organized by the IFLA Rare Books and Manuscripts Section, Munich, 19–21 August 2009*, ed. Bettina Wagner and Marcia Reed (Berlin: De Gruyter Saur, 2010), 321–328.

7 Sidney Lee, *Shakespeares Comedies, Histories, & Tragedies: A Supplement to the Reproduction in Facsimile of the First Folio Edition (1623) from the Chatsworth Copy in the Possession of the Duke of Devonshire, K.G., Containing a Census of Extant Copies, with Some Account of their History and Condition* (Oxford, UK: Clarendon Press, 1902); Seymour De Ricci, *A Census of Caxtons* (London, UK: Printed for the Bibliographical Society at the Oxford University Press, 1909).

8 Owen Gingerich, *An Annotated Census of Copernicus' De Revolutionibus (Nuremberg, 1543 and Basel, 1566)* (Leiden: Brill, 2002).

9 Daniel Margócsy, Mark Somos, and Stephen N. Joffe. *The Fabrica of Andreas Vesalius: A Worldwide Descriptive Census, Ownership, and Annotations of the 1543 and 1555 Editions* (Leiden: Brill, 2018).

10 Marieke van Delft, "Exemplaren Wereldwijd van *Metamorphosis Insectorum Surinamensium* 1705 = Worldwide Copies of *Metamorphosis Insectorum Surinamensium* 1705," in *Metamorphosis Insectorum Surinamensium: Verandering der Surinaamsche Insecten = Transformation of the Surinamese Insects: 1705* (Tielt: Lannoo, 2016), 187–189.

11 Paul Hoftijzer, "Zo Vergaat de Roem: Het Einde van de Officina Hackiana," in *Van Pen tot Laser: 31 Opstellen over Boek en Schrift Aangeboden aan Ernst Braches*, ed. T. Croiset van Uchelen and H. van Goinga (Amsterdam: De Buitenkant, 1996), 160–167.

12 Databases consulted were: "Book Sales Catalogues Online (BSCO)," *Brill*, accessed 20 May 2022, https://brill.com/view/db/bsco; "Material Evidence in Incunabula (MEI)," *CERL*, accessed 20 May 2022, https://data.cerl.org/mei/_search; "KVK - Karlsruher Virtueller Katalog," *Karlsruher Institut für Technologie*, accessed 20 May 2022, https://kvk.bibliothek.kit.edu/; "Short-Title Catalogue Netherlands (STCN)," *KB Nationale Bibliotheek*, accessed 20 May 2022, https://www.kb.nl/over-ons/diensten/stcn; "Language of Bindings Thesaurus (LoB)," University of the Arts London, *Ligatus*, accessed 20 May 2022, https://www.ligatus.org.uk/lob/; "Worldcat," *OCLC*, accessed 20 May 2022, http://www.worldcat.org/.

13 Hoftijzer, "Zo Vergaat de Roem," 160.

14 Alphonse Willems, *Les Elzevier: Histoire et Annales Typographiques* (Bruxelles: G.A. van Trigt, 1880), CCXII.

15 Sachiko Kusukawa, *Picturing the Book of Nature: Image, Text, and Argument in Sixteenth-Century Human Anatomy and Medical Botany* (Chicago, IL: University of Chicago Press, 2012), 50–61.

16 Rijksmuseum, object numbers: RP-P-OB-200.099, RP-P-OB-84.391, RP-P-OB-84.392, and RP-P-OB-200.099. We thank Erik Hinterding for bringing this to our attention.

17 Paul Begheyn, "De Elzeviers en de Jezuïeten," in *Boekverkopers van Europa: Het 17de- Eeuwse Nederlandse Uitgevershuis Elzevier*, ed. Berry P.M. Dongelmans and Paul Hoftijzer (Zutphen: Walburg Pers, 2000), 59–76.

18 Jan Storm van Leeuwen, "De Introductie van het Stempelen à Petits Fers en de Nederlandse Boekband tussen ca. 1620 en ca. 1665," in *Opstellen over de Koninklijke Bibliotheek en Andere Studies* (Hilversum: Verloren, 1986), 262–263.
19 Whitehead and Boeseman, *Dutch 17th Century Brazil*, 28.
20 Truusje Goedings, *'Afsetters en Meester-afsetters': De Kunst van het Kleuren 1480–1720* ([Nijmegen]: Vantilt, 2015), 19.
21 *Catalogus van een Menighte Treffelijcke Boecken [...] Naergelaten by Wijlen Frans Koerten [...]* (Amsterdam: Jacob Lescailje, 1668).
22 Abraham J. van der Aa, *Biographisch Woordenboek der Nederlanden, Bevattende Levensbeschrijvingen van Zoodanige personen, die zich op Eenigerlei Wijze in Ons Vaderland Hebben Vermaard Gemaakt* IX (Haarlem: Brederode, 1860), 152–159.
23 Goedings, *Afsetters*, 25.
24 Delsaerdt, "De Verzegelde Kisten," 146–147.
25 Dominique Varry, "Revolutionary Seizures and Their Consequences for French Library History," in *Lost Libraries: The Destruction of Great Book Collections since Antiquity*, ed. James Raven (Basingstoke, UK: Palgrave Macmillan, 2004), 182.
26 Friedrich Buchmayr, "Secularization and Monastic Libraries in Austria," in Raven, *Lost Libraries*, 145.
27 Bert Sliggers, *Herkomst: Boudewijn Büch* (Amsterdam: De Arbeiderspers, 2005), 89–91.
28 Richard I. Johnson, "The Rise and Fall of the Boston Society of Natural History," *Northeastern Naturalist* 11, no. 1 (2004): 81–108.
29 Boeseman et al., "Seventeenth Century Drawings."
30 Marinus Boeseman, "A Hidden Early Source of Information on North-Eastern Brazilian Zoology," *Zoologische Mededelingen* 68 (1994): 124.
31 Alex Alsemgeest and Charles Fransen, *In Krabbengang door Kreeftenboeken: De Bibliotheca Carcinologica L.B. Holthuis* (Leiden: Naturalis Biodiversity Center, 2016).

4 Searching for Copaiba

Tracing the Quest for a Wound-Healing Oil by Early Explorers in Brazil

Tinde van Andel, Mariana Françozo, and Mireia Alcantara Rodriguez[1]

Introduction

When Count Johan Maurits of Nassau-Siegen was appointed governor of Dutch Brazil in 1636, he commissioned a group of scientists and artists to document the flora, fauna, and cultures in this new Dutch colony.[2] The count's support of natural history, astronomy, geography, and scientific and ethnographic illustration during his governorship was highly unusual and distinguished him from other colonial administrators and military leaders in the seventeenth century.[3] The *Historia Naturalis Brasiliae* (henceforth HNB), with its beautiful and accurate illustrations of plant and animal life, was one of the first comprehensive publications of South American natural history and had a substantial influence as a reference work among European scholars.[4]

The identification of the plants described in the HNB is difficult due to the crude woodcut illustrations and the early seventeenth-century Latin descriptions,[5] but is greatly facilitated by Marcgraf's herbarium,[6] which is praised as the first to hold dried plant specimens from tropical America.[7] Recent studies on sixteenth-century herbaria, however, have discovered several older Neotropical specimens, grown in European botanical gardens from seeds brought from the Americas around the 1560s.[8] These Renaissance book herbaria, however, only contain a handful of cultivated plants (e.g., tomato and chili pepper), without any notes on geographic origin or uses, while Marcgraf's herbarium contains 145 species of mainly wild Brazilian plants, of which 103 are also described by Marcgraf and Piso in the original HNB and/or by Piso in what became known as the second edition of the treatise.[9]

For many European doctors and pharmacists, the HNB offered a first introduction to various Brazilian medicinal plants and their effect on the human body.[10] Species such as ipecacuanha (*Carapichea ipecacuanha* (Brot.) L. Andersson), used against dysentery; the insect-resistant cabuberiba balm (from *Myroxylon balsamum* (L.) Harms); the laxative seeds of pinhones (*Jatropha curcas* L.); and the wound-healing copaiba balsam

DOI: 10.4324/9781003362920-5

(*Copaifera* spp.) quickly spread their fame and were shipped to Europe in large quantities toward the end of the seventeenth century.[11] Instead of only copying what was already known from letters and books by previous explorers, missionaries, and colonial authorities, Piso and Marcgraf had the chance to make their own observations. Marcgraf took part in organized expeditions into the Brazilian wilderness and probably used the opportunity to collect specimens, while Piso experimented with medicinal plants on himself or on local inhabitants.[12] The HNB has been repeatedly praised as the most important contribution to the science of natural history since Aristoteles and Pliny.[13] Carl Linnaeus, the godfather of botany, considered the scientific descriptions and illustrations in the HNB of such high quality that he used several of them for the tenth edition of his taxonomic masterpiece *Systema Naturae*.[14]

Marcgraf and Piso, however, were not the first to document Brazilian herbal medicine. Portuguese Jesuit missionaries were engaged in substantial bioprospecting activities since the 1550s, collecting knowledge on local herbal medicine from Indigenous healers to address the health problems of Portuguese settlers in the South American tropics. Although the Jesuits' ethnopharmacological work was passed on to Portuguese physicians, surgeons, pharmacists, and colonial officials, many of these early writings on traditional remedies and their natural ingredients only survive as unpublished manuscripts.[15] Portugal did not send out state-sponsored scientific expeditions to systematically study and record the flora and fauna of their overseas territories until the late eighteenth century.[16]

This chapter focuses on early reports of one important Brazilian medicinal product: copaiba balsam. Although the HNB was applauded for providing the first explicit description and illustration of one of the trees yielding this oily exudate, the exact species of *Copaifera* that yielded this highly valued medicine remained shrouded in mystery long after 1648.[17] Although the plant species in the HNB have been subjected to botanical revision, in which the copaiba tree was identified as *Copaifera officinalis* L., a recent revision of the useful plants described by Marcgraf and Piso indicates that many of these identifications were outdated or inaccurate.[18]

We list the earliest reports on this herbal product in Brazil and trace attempts to describe the plant species that yield the copaiba balsam, its uses, and extraction method. We explain how Marcgraf's unexpected early death led to the erroneous combination of his encrypted information on copaiba with other descriptions and illustrations, leading to a confusion in taxonomy, local names, and interpretations, which lasted for centuries. We show that by studying the original texts and illustrations from diverse sources in the scientific entourage of Count Johan Maurits, of which several only recently became available to scientists,[19] we can finally link the wound-healing oil described in the HNB to two species of *Copaifera*.

Methods

In 2014, we conducted an ethnobotanical research on the useful plants described in the HNB and those in the so-called second edition of this treatise, in fact a somewhat different book published by Willem Piso under the title *De Indiae utriusque Re Naturali et Medica*, both held in the Rare Book Room of the library of Naturalis Biodiversity Center in Leiden, the Netherlands.[20] We consulted the original Latin copies of the two aforementioned tomes and the Portuguese translations of both works.[21] We also studied Marcgraf's original herbarium in the Botanical Museum of the University of Copenhagen and compared its specimens with the earlier revision by Andrade de Lima et al.,[22] the identifications of the plants in the HNB by the Brazilian botanist B.J. Pickel,[23] the online checklist of Brazilian flora,[24] the digital database on Brazilian herbarium specimens SpeciesLink,[25] and the Brazilian herbarium collections in the herbaria of Copenhagen, Missouri Botanical Garden,[26] and Naturalis.[27] For the distribution of the different *Copaifera* species, we consulted the Global Biodiversity Information Facility (GBIF) database.[28] We updated the scientific names by using The Plant List.[29]

For the present paper, we focused on historical attempts to describe the wound-healing copaiba balsam and the search for the tree that produced this valuable product in Brazil. We built on a previous article in Dutch on the confusion around copaiba balsam due to the erroneous combination of plant descriptions and drawings by Johannes de Laet, the editor of the HNB.[30] To verify what knowledge was already available before the HNB was published in 1648, we added information on the earliest reports on Brazilian copaiba balsam (1550–1647) from historical texts mentioning words in Tupi-related Indigenous languages listed by Cunha.[31]

We did not review seventeenth- and eighteenth-century reports on copaiba oil found outside Brazil, such as those from the Guianas or the Caribbean islands, as these likely describe different species of *Copaifera* than the Brazilian sources – or no *Copaifera* at all.[32] To trace when copaiba was first mentioned in Dutch pharmacopoeias or trade documents, we queried the Time Capsule database, an online search engine that links several datasets relating to the early modern history of medicinal plants in the Low Countries between 1550 and 1850.[33]

We also consulted the entries on copaiba or similarly named plant species in a manuscript containing notes by Marcgraf and passed on to De Laet, presently kept at the British Library.[34] Finally, we examined the digital images of several oil paintings of plants that had not been taxonomically identified, made during the 1630s–1640s by artists in the circle of Johan Maurits and currently kept in the *Libri Picturati* collection housed by the Jagiellonian Library in Kraków, Poland.

Results

The first Europeans arrived in Brazil in 1500, but permanent settlement began only a few decades later, for instance near São Paulo only in 1532. Somewhere between 1513 and 1521, in one of the first accounts of explorations in South America, the Italian historian Petrus Martyr of Anghiera (1457–1526) wrote in a letter to Pope Leo X about a resin-producing tree named "copei."[35] He was probably the first European to mention copaiba balsam (Table 4.1). On 31 May 1560, the Jesuit priest José de Anchieta (1534–1597) wrote a letter to his Spanish colleague Diego Laynes (1512–1565), in which he mentioned a tree that supplied a sweet balm that was produced by incisions with knives or axes in the bark. It reminded him of a Swedish distillate and cured wounds so quickly that no scars remained. Anchieta had used it himself. This unnamed tree was later connected to a tree described as "cupaigba" by another Jesuit priest, Fernão Cardim, between 1583 and 1601 and a tree named "copaíba" by Soares de Sousa and identified by Hoehne as *Copaifera officinalis* L.[36]

The chronicler Pêro de Magalhães Gândavo (1540–1580) was the first to mention that animals also know the healing properties of copaiba balsam, which was later confirmed by the Jesuit priest Fernão Cardim (1548?–1625). The latter wrote a more detailed account of copaiba during his stay in Pernambuco between 1583 and 1601, which was only published centuries later.[38] Cardim thought that "cupaigba" was a fig tree ("figueira"), but also described the clear, oily exudate that was used for wound healing and added that it was inflammable and could be used as a light source. This is hardly ever the case for the white, non-transparent exudate of *Ficus* trees. Cardim considered the wood to be worthless. Typically, exactly the same description, including the inaccurate identification as a fig tree, was attributed to the Portuguese monk Manoel Tristão of the convent of Bahia, whose account on this oil under the name "cupayba" was published by Samuel Purchas and often considered as the first or second written account on copaiba oil.[39] Ambrósio Fernandes Brandão (1555–1618), sugar mill owner in Paraíba, reported that wounded soldiers were treated with the oil,[40] a use that is not mentioned afterwards anymore (Table 4.1). In his treatise on the country and people of Brazil from 1587, Portuguese farmer, landowner, and scholar Gabriel Soares de Sousa also gave a detailed description of "the most holy oil," but considered the fruits to be inedible.[41]

Around 1594, the Jesuit priest Francisco Soares described the copaiba as a tall and thick tree with very hard wood that yielded a unique, wound-healing oil that had its best quality in summer. When he was on a ship, he cut off his fingertip (which fell overboard) and rubbed his wound with copaiba oil. The injury did not get infected and left only a thin white scar. He tried to convince the ship surgeons to use this oil as well. He mentioned a request to the bishop of Brazil for a license to commercially produce and export the oil. He ends his praise with the suggestion that "there are many things

Table 4.1 Historical accounts that mention copaiba balsam from Brazil, ordered chronologically.[37]

Author (year when copaiba is mentioned)	Botanical description	Image	Local name	Use description (translated text)
Petrus Martyr of Anghiera (1513–1521?)	?	no	copei	Resiniferous tree?
Father José de Anchieta (31 May 1560)	"tree"	no	not given	Resin harvested by incisions, sweet scent, wound healing, prevent scars.
Pêro de Magalhães Gândavo (1576)	"tree in Pernambuco"	no	copahíbas	Resin harvested from bark, wound healing, eases pain, wounded animals also use it.
Jean de Léry (1578)	tree, looks like walnut tree?	no	copa-u	Wood used for furniture.
Anonymous Jesuit priest (c. 1580)	?	no	copaíba?	?
Gabriel Soares de Sousa (1587)	large tree, not very hard wood,	no	copaíba	Fruit inedible, oil used in lamps, harvested with axes, runs into bottles, good smell, applied on wounds and burns, prevents scars, for colds, stomach aches, most holy oil, used in households. Wood used to make wooden shields.
Carolus Clusius (1605)	A liquid or gum brought from the West Indies	no	copal-yva	Strongly recommended and I understand that it is very useful for curing fresh wounds.
Father Fernão Cardim (1583–1601)	Common, tall, straight and thick fig tree	no	cupaigba	Contains abundant oil, sometimes more than a quarter. Oil is very bright, olive color. Highly esteemed for wounds, takes away every sign, also used for candles, burns well. Animals rub against the bark. Wood of no value.
Father Francisco Soares, c. 1594	tall and thick tree	no	copajba	Hardwood, wound-healing oil, own experiments, prevents scars, internally as laxative, against swellings.
Father Simão Travaços, c. 1596	trees that give the balm from Ilheos and Espírito Santo, are the best in the world	no	not given	Trees that [when] cut they give much oil from the cup that has great virtues for wounds, and discharges.

(Continued)

Table 4.1 Historical accounts that mention copaiba balsam from Brazil, ordered chronologically.[37] (*Continued*)

Author (year when copaiba is mentioned)	Botanical description	Image	Local name	Use description (translated text)
Ambrósio Fernandes Brandão (1618)	"plants found in the southern provinces"	no	copaiba	Wounded soldiers are readily healed with native copaíba, a balsam confected from plants.
Manoel Tristão (1625)	a fig tree, commonly very high, straight and big		cupayba	It has much oil; to get it they cut the tree in the middle, where it comes out in great abundance, sometimes more than a quarter; very clear colored oil; much used for wounds, takes away all the scars. Also for lights and burns well. Animals rub themselves to the trunk. Wood not used.
Johannes De Laet (1625, 1640)	Very common tree, similar to fig tree, high, big and straight		cupayba, copal-yua	Contains much oil, obtained by cutting the bark, heals wounds, prevents scars, also as lamp oil. Wood is not useful.
Amsterdam Pharmacopeia (1643)	no	no	balsam Copa-ivae	No
Adriaen van der Dussen (1637)	"famous tree"	no	copaiba	Sweet-scented balsam though incision in bark, miraculous wound and scar healing, used by animals bitten by snakes.
Georg Marcgraf (1648)	detailed description of leaves, flowers, fruits.	yes, fruit only	Copaiba Brasiliensi bus	Detailed description of oil properties, harvest methods, medicinal recipes and application: wounds, nerves, diarrhea, dysentery; fruit pulp edible.
Willem Pies (1648 and 1658)	Detailed description of wood, leaves, bark, fruits, fruiting period, distribution	yes, but forged image	Copaiba, Copaliba	Detailed description of harvest methods, recipes, properties and application of oil: against "espinela", severe diarrhea, dysentery, gonorrhea, wounds, ulcers, nerves, breast disorders, abdominal colic, menstruation, flatulence, mosquito and snake bites. Fruit eaten by monkeys and humans; wood used for boards.

that could be written [about the oil that it could be] a book." Soares' manuscript, however, was only published three centuries after his death.[42]

Even the Dutch had written about copaiba before the HNB was published. Adriaen van der Dussen, employee of the West India Company (WIC), noted in 1637: "Among their most famous trees is copaíba, of which the sweet-smelling balsam comes from an incision in the bark, healing wounds and removing scars with a miraculous force. The tree can be recognized by the damage done on the bark by wild forest animals, which know by natural instinct to rub their skin against its bark when they are bitten by snakes." His account was published as part of historian Caspar Barlaeus' *Rerum per Octennium...* in 1647.[43]

Almost 20 years before the HNB was published, copaiba oil was apparently already shipped to Europe in such quantities that it was mentioned in the Amsterdam Pharmacopeia of 1630.[44] This name was also used in the Amsterdam Pharmacopeia of 1643 as "balsam Copa-ivae." This name was later changed into balsamus capivi or copaiba.[45]

Johannes de Laet, one of the founding directors of the WIC, had already written a description of the "New" World in 1625, which was first published in Dutch, then in Latin, and finally in French, with new additions in each edition.[46] In De Laet's reference to copaiba oil, he cited the work of the French botanist Carolus Clusius, who had translated the book *Tractado de las Drogas y Medicinas de la Indias Orientales* by the Portuguese doctor Cristóbal Acosta.[47] Acosta had received several bottles of copaiba oil from his overseas friends. De Laet's description of copaiba balsam, however, echoes the earlier descriptions of the Jesuits rather than Clusius' description (Table 4.1).

To inform Johan Maurits on the situation in Dutch Brazil, De Laet compiled a handwritten guide for the new colony,[48] in which he gave a detailed account of the geography of the area, as well as suggestions where and how to attack the Portuguese and what goods could be obtained from the local inhabitants in specific areas. De Laet described that in "Marannon," the Indigenous people were willing to trade cotton, food, dyes, silver, and "a balsam oil that they call uwijraca-andugh, growing on the copaíba tree."[49] Typically, this Indigenous name does not appear anymore in the HNB, in which only the Tupi name copaíba is given.[50] It is likely that Marcgraf and Piso used De Laet's early work to compile a "wish list" of useful Brazilian plants that needed professional scientific descriptions.

Marcgraf's Description: Scarlet Wood with a Turpentine-Like Oil

On pages 130 and 131 of the HNB, Marcgraf described "Copaiba Brasiliensibus" as a tree with mostly deep scarlet wood, hard as beech wood, which was sawed into wide planks for diverse applications (Figure 4.1).

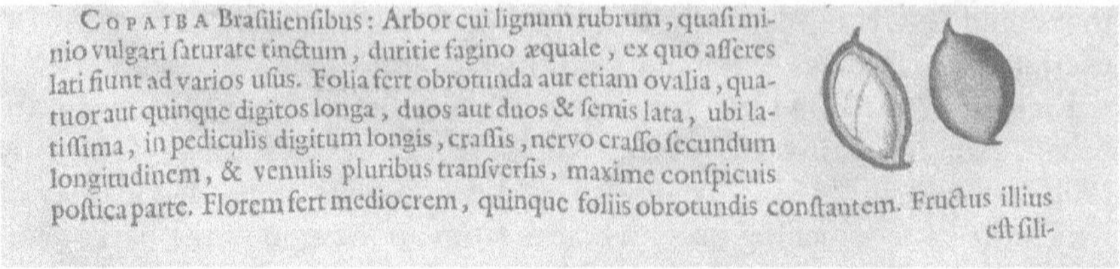

CoPAIBA Brafilienfibus: Arbor cui lignum rubrum, quafi mi-
nio vulgari faturate tinctum, duritie fagino æquale, ex quo afferes
lati fiunt ad varios ufus. Folia fert obrotunda aut etiam ovalia, quatuor aut quinque digitos longa, duos aut duos & femis lata, ubi la-
tiffima, in pediculis digitum longis, craffis, nervo craffo fecundum
longitudinem, & venulis pluribus tranfverfis, maxime confpicuis
poftica parte. Florem fert mediocrem, quinque foliis obrotundis conftantem. Fructus illius
                                                                                                          eft fili-

Figure 4.1 Description of "Copaiba Brasiliensibus" by Marcgraf with the woodcut
image of the opened and closed fruit of *Copaifera* in *Historia Naturalis
Brasiliae* (Piso and Marcgraf, 1648: part II, 130). Leiden University
Libraries (copy 1407 B 3).

The leaves of the tree were round or oval, four or five fingers long, and two
to two-and-a-half fingers wide, on a stalk of a finger long, with thick sec-
ondary longitudinal veins and many transverse veins, the most strikingly
visible on the back. The tree had small flowers with five roundish petals.
The fruits were small, brown, and round pods, the size of a finger, and easy
to open by hand. They contained a seed the size and shape of a hazelnut,
covered with a black, membrane-like skin embedded in a little yellow pulp,
with a scent of crushed peas. This soft, tough pulp had an unclear, watery
taste, but was nevertheless eaten. Ripe pods all fell from the branches at
the same time. The tree produced a remarkable oil or balm with a resinous
odor and drops that resembled turpentine oil in taste and consistency. The
oil was harvested by drilling a hole at the base of the trunk into the sap-
wood and placing a small container under it. About four cups of oil could
be harvested within an hour. Because the oil continued to flow, the gap was
often closed from dawn to dusk, which undoubtedly pointed to the importance of the oil. A small amount of heated oil was applied on fresh wounds,
after which they stopped bleeding and healed quickly. Three or four drops
of oil were mixed with a fresh egg and taken two or three times on one
morning against nerves. The oil helped to cure dysentery and other forms
of diarrhea. The oil was considered warm and dry in the second degree.
Unfortunately, the first and only illustration of the copaiba included in the
HNB is a woodcut image of an opened and a closed fruit. Marcgraf probably picked them up from the forest floor, as he wrote that the ripe pods fell
massively from the trees. Although he had seen the leaves and flowers of the
copaiba tree, they are missing from his herbarium.

Piso's Description: An Oil with Remarkable Uses

Piso mentioned copaiba several times in his work. In his chapters of the
HNB, bundled together and called *De Medicina Brasiliensi*, there is a
lengthy description of the copaiba tree. Piso wrote that the "province of
Brazil" produced various balms, of which copaiba is the most important.
Copaiba was the name of the tree from which it came: a tall tree with gray

bark that grew in the wild. The leaves were half a foot long and consisted of larger and smaller leaves that faced each other, with fine veins and a pointed tip. Young leaves were rusty brown. At the end of the branches, among the leaves, the flower clusters were found. These were followed by fruits with the size and shape of bay berries, first green, then black after ripening, with little, slightly sweet-tasting flesh. The fruits contain an oval, hard seed, thicker than that of the wild plum, covered with black skin that was easy to remove and containing a white core with a floury taste, but not edible. The fruit ripened in April and was eaten by the Brazilians, who consumed the juice and spat out the black skin. Monkeys also enjoyed the fruits very much. Piso recalled that "in the month of June I collected fruits that were already half germinated, and I ordered the earth to increase the yield."[51]

Earlier in the HNB, Piso had already mentioned several medicinal applications of copaiba oil. To heal "espinela" (a pain near the solar plexus),[52] some drops of copaiba oil were dissolved in a generous amount of wine and taken internally, and for an external poultice on the stomach, the oil was mixed with the exudate of icicariba (*Protium heptaphyllum* (Aubl.) Marchand), cabureiba balsam (*Myroxylon balsamum* (L.) Harms), egg yolk, and saffron.[53] Apart from the vomiting-inducing ipecacuanha roots (*Carapichea ipecacuanha*) and the strongly laxative seeds of pinhones (*Jatropha curcas*), Piso recommended for severe diarrhea and dysentery the oral intake of some drops of copaiba oil, dissolved in sugar and beaten egg. The rectal administration of this mixture was also prescribed to comfort inflammation of the anus.[54]

For "the virulent gonorrhea," Piso mentioned that once the disease was defeated, most experts limited themselves to prescribe astringents, consolidating and drying agents.[55] He recommended copaiba balsam, dissolved in sugar or olive oil, as the best medicine, either taken orally or injected in the penis. Wounds and ulcers were healed with the scented balsams cabureiba (*Myroxylon balsamum*) and copaiba: they did not only stop the bleeding, but, applied internally and externally, also fortified the nerves. The two balsams were considered to have the same quality.[56] He also described that the brave men who travel the backwoods or dense forests of Brazil, where the sea breezes barely arrived, anointed their naked members with the balsams of copaiba and cabureiba.[57]

Piso also provided a description of the harvesting practices of copaiba oil: "the tree is rich in fragrant liquid. One makes cuts in the bark of this huge tree, preferably in the period up to the full moon, so that a large amount of oil droplets come out. In three hours, 12 libras flow out without difficulty. If no oil flows out, close the hole in the bark with clay or wax. After two weeks the yield will be enough to compensate for the delay. This tree is not so much found in the Pernambuco prefecture but especially on the island of Maranhon. An abundance of [copaiba] balm is growing here, which is why we can afford the supply of this balm."[58]

About the medicinal use of the oil, Piso stated: "not only does the oil have an amazing cleansing and stabilizing capacity, it is primarily used to heal wounds, mosquito and snake bites, and to remove scars. Not only the locals, I myself have also noticed the remarkable usefulness of this oil. He is not as sweet-smelling as required by Maffeus. [The oil is] warm in the second degree, thick, very greasy and resinous. In drops administered orally, it relieves breast disorders, abdominal diseases and cold colic. The oil provides vital strength, it stops women's periods, flatulence, and gonorrhea. A similar success against this evil can be achieved by means of a syringe in the anus or in the penis with [copaiba drops] dissolved in [an extract of] plantain water (an extract of *Plantago major* L.) or rose oil."

Piso did not add a drawing of the copaiba tree. It was probably also difficult for him to get the leaves because the tree, as he wrote it himself, did not occur in the neighborhood of Mauritsstad, but in the Maranhão area. At one moment in June, however, he obtained germinating seeds and probably planted them in the garden of Vrijburg, the walled garden of Johan Maurits where many plant species were grown and wild animals were kept in cages for further study.[59] We do not know whether the copaiba seedlings grew successfully.

Secrecy and Distrust in Dutch Brazil

Around 1644, Marcgraf travelled to Angola to map the Dutch possessions for the WIC, but he died from yellow fever shortly after arriving in Central Africa.[60] Since he had not yet published anything when he left Brazil, Marcgraf had entrusted his botanical collections, manuscripts, and drawings of plants and animals to Johan Maurits before he left. He had written his notes in a secret code, probably out of fear of plagiarism by Piso. Although they initially worked closely together, their relationship was later characterized by jealousy and distrust.[61]

Johannes de Laet managed to decipher the secret code and edited Marcgraf's manuscripts together with Piso's notes on indigenous diseases and medicinal plants and published them together in the HNB.[62] The entry on copaiba in De Laet's British Library manuscript does not differ substantially from the final version in the HNB.[63] There is no woodcut proof attached to the opposite page, as is the case on other pages of the manuscript, and no reference is made to an image elsewhere (Figure 4.2).

Marcgraf's lack of confidence in Piso's integrity, however, proved to be correct. In 1658, after De Laet's death, Piso published *De Indiae utriusque...* as sole author, in which he incorporated Marcgraf's figures and descriptions into his own text, without mentioning him as an author, for which Linnaeus accused him of plagiarism.[64] Linking his own collected information about medicinal plants to Marcgraf's botanical descriptions, Piso made a number of mistakes in the transcription of the text, the retouching of illustrations, and the identification of species. This plagiarism, and

Figure 4.2 Entry on "Copaiba" from Johannes De Laet's manuscript (De Laet, n.d.: f. 68). © The British Library Board (Sloane Ms. 1554).

the consequent confusion, has caused major headaches to (ethno-) botanists in their interpretation of the historic descriptions of plants and their uses in seventeenth-century Brazil.[65]

Copaifera spp. (Leguminosae) versus *Clusia nemorosa* (Clusiaceae)

In his 1658 "version" of the HNB, Piso again devoted a paragraph to the copaiba: "most Americans call all scented resins and gums copal, although there are various species with different names. Therefore, all resin-bearing trees in Brazil are simply called copaliba or copaíba. In the dense forests of the interior this often happens [with trees] whose wood is red as vermilion and so hard that it is used to make wide boards." Piso continued with a description of the copaiba. This time, however, he did not mention compound leaves but suddenly described a flower with five rounded petals. The description of the dark pod with the watery, edible flesh and the extraction of the richly scented oil is the same as in earlier versions of his own text and that of Marcgraf. For the medicinal uses, Piso added that the healing power of the oil was proven again during Jewish circumcisions: "after treatment with copaíba oil, the blood flows very limitedly from this cruel wound. Previously it was difficult to effectively heal wounds, now this oil works without any problems."[66]

Piso's description is accompanied by a woodcut image of the copaiba tree, but this time the image contains leaves and flowers (Figure 4.3). The leaves are not compound, as is usually the case with members of the Leguminosae family and always the case in the genus *Copaifera*. The flower strongly resembles that of *Clusia nemorosa*, a species described and depicted by Marcgraf under the local name "coapoiba" or "pao gamelo" in Portuguese "of which several species exist, two of which will be described here."[67] Marcgraf mentioned that the leaves of the first species of "coapoiba" had almost invisible veins and produced a white exudate when they were cut off. The flowers were

Cᴏᴘᴀɪ́ʙA.

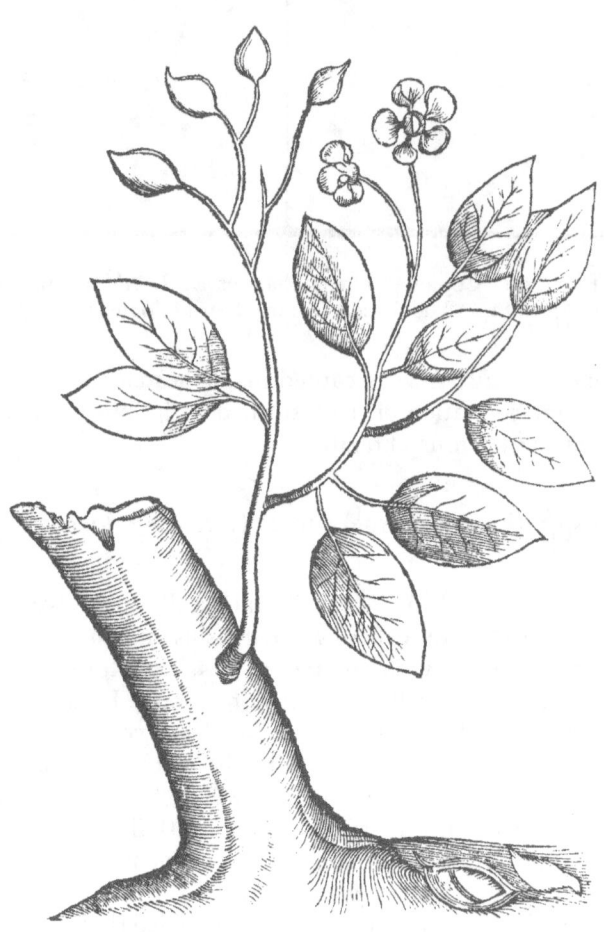

Figure 4.3 Woodcut image from Piso's *De Indiae utriusque* (Piso, 1658: 118). A combination of the flowers of *Clusia nemorosa* (branch on the right side) and the fruits of *Copaifera* (branch on the left side) and leaves of an unknown origin. Republished by Elsevier B.V. 2013.

as large as roses, the petals white with "soft pink like toenails and a navel in the middle, in the shape of a sticky yellow bulb." The fruit contained a yellow exudate split open lengthwise and rows of seeds in a red pulp. The bark and the marrow could be easily separated from the wood. In the seventeenth century, the term "gamelo" referred to a wooden bowl, used as a container, which was probably made from the wood of this species.[68]

Marcgraf's entry on "coapoiba" is a very adequate description for a species of the genus *Clusia*, which is depicted in Marcgraf's woodcut image on page 131, directly after his description of *Copaifera*, and represented with two specimens in his herbarium that were identified by us as *Clusia nemorosa* G. Meyer (Figure 4.4A–C). Marcgraf finally noted that the fruit of "coapoiba" was dry with no pronounced taste. Although he heard that some people ate it, he found it worthless. He did not mention any medicinal use.

What Went Wrong with Copaiba?

Piso, who scornfully wrote that "Americans confused all trees with fragrant resin," seemed to be making the same mistake himself. Why were two very different species forged together in one woodcut image? Did he think that such an economically important tree as *Copaifera* deserved a complete illustration? Piso's description of copaiba is placed in his fourth chapter in the HNB, after sugar (chapter 1), cassava (chapter 2), and wild honey (chapter 3). Copaiba balsam must have been of great economic importance because of its multiple medicinal properties, widespread use, and trade. Although there are major differences in flowers and fruits, the local names copaíba and coapoiba are indeed quite similar, and both trees produce exudate, although the sticky, pale yellow or white latex excreted by *Clusia* species differs substantially from the colorless oil of *Copaifera* trees. The origin of the name copaíba is found in the Indigenous Tupi language, in which it means "deposit tree," referring to the amount of oil it produces.[69]

According to the Brazilian botanist Pickel, the woodcut of copaiba in Piso was a "fantasy" and a "bluff."[70] The faulty woodcut could also have been made by De Laet, who produced several missing illustrations based on Marcgraf's herbarium to include in the HNB. It is often unclear which descriptions the illustrations belong to, possibly because of De Laet's limited botanical knowledge or his problems with deciphering Marcgraf's secret code.

The exact species of *Copaifera* that was described and depicted by Marcgraf and Piso is difficult to trace from the published texts of the HNB. When the French botanist Von Jacquin found a flowering *Copaifera* tree on Martinique in 1760, he considered it to be identical to the species described by Marcgraf and named it *Copaiva officinalis* Jacq., literally "medicinal copaiba," even though the tree had four instead of five petals.[71] Linnaeus based his description of the species *Copaifera officinalis* (Jacq.) L. (literally "medicinal copaiba-bearing [tree]") on the specimen collected by Jacquin

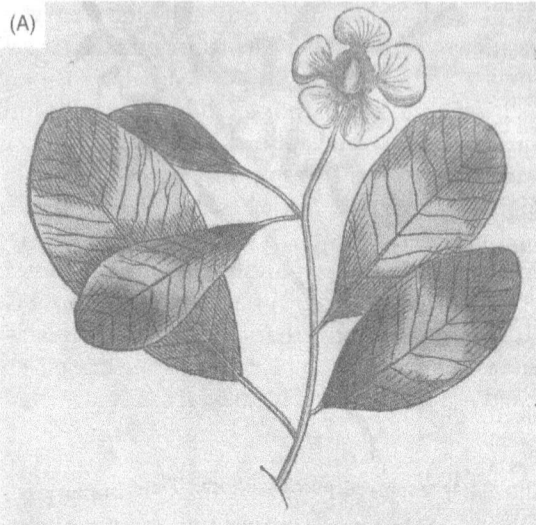

oblongæ. Cortex hujus arboris detractus glutinofus est, & derafa exteriori cute fufca, feu recens feu ficcus nfurpatur, egregie Saponis vicem implet & tuto adhibetur ad omnia ad quæ Sapo Hifpanicus.

Longe præstat fructui *Sabaon*, ille enim acrimonia fua nocet vestibus, hic autem nequaquam

Nafcitur ubique in Brafilia quidem maxima copia.

Coapoiba Brafilienfibus, *Pao Gamelo* Lufitanis; hujus Arboris aliquot reperiuntur fpecies; *quarum duas Auctor ita defcribit.* Prima fpecies in Fagi altitudinem & figuram excrefcit, cortice cinereo cui aliquid fufci admixtum, instar undulati panni. Folia habet folida, oblonga, inferius dilute virentia fuperius faturatiora & fplendentia, infigni nervo fecundum longitudinem, at nullis pene venis confpicuis

R 2

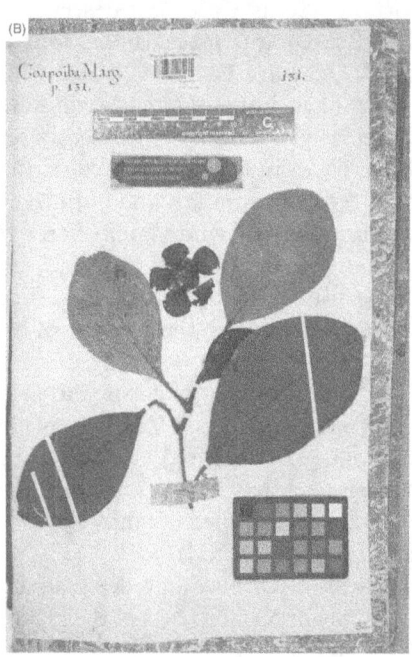

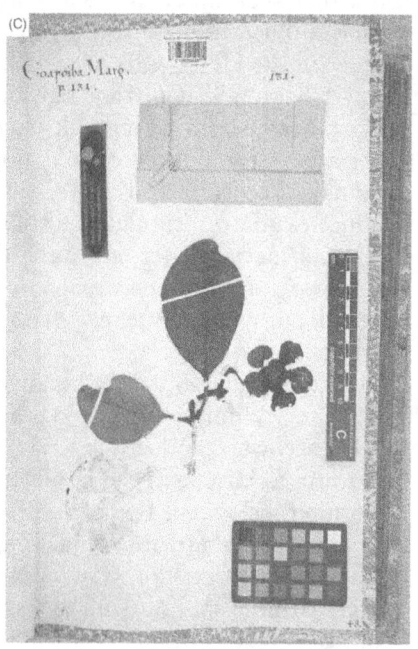

Figure 4.4 (A) Marcgraf's woodcut image of "Coapoiba" (*Clusia nemorosa*) in *Historia Naturalis Brasiliae* (Piso and Marcgraf, 1648: part II, 131). Leiden University libraries (copy 1407 b 3). (B) and (C) Marcgraf's herbarium Collections of *C. nemorosa* that were used as models for this illustration (The Marcgrave Herbarium, 1638–1644: 32, 48). Image published with permission from Herbarium C, Natural History Museum of Denmark.

but confirmed that Marcgraf's copaíba and Piso's coapoiba belonged to the same species.[72] This erroneous identification was later copied by the French botanist Aublet in his influential work on the flora of French Guiana.[73] In a later edition of the *Species Plantarum*, the reference to Piso's description was corrected to the copaiba in the original edition of the HNB.[74] In 1949, Pickel identified the copaiba tree described in the HNB as *C. officinalis*, but the Brazilian physician and parasitologist Pirajá da Silva identified the species in the HNB as *C. langsdorffii* Desf,[75] although it is unclear on what morphological characters they based their decisions.

Theatri Rerum Naturalium Brasiliae

In 1652, Johan Maurits gifted hundreds of unbound oil paintings and drawings of Brazilian plants and animals to Friedrich Wilhelm, Elector of Brandenburg, which were later reorganized and bound by the Elector's physician, Christian Mentzel, into four volumes: the *Theatri Rerum Naturalium Brasiliae*, or *Libri Picturati* A 32–35.[76] This collection, currently housed by the Jagiellonian Library in Kraków, has not been examined by botanists for centuries, but has recently been digitized. Some of these illustrations served as the basis for the woodcut illustrations in the two editions of the HNB.[77] In the fourth volume (A 35), dated 1662 and containing 171 illustrations of plants glued on sheets of paper, several pages are left blank. Folio 77 was intended to contain a painting of "Copaiba P. p. 118. Coapoiba. Marg. p. 130" (Figure 4.5). Did Mentzel have to wait for the missing drawings because they were left at the publishers? Or was he confused about the similarity of the local names and did not know which image to include: Marcgraf's *Copaifera* fruits (Figure 4.1), the forged image (Figure 4.3), or the image of *Clusia nemorosa* (Figure 4.4A)?

Mysterious Paintings Identified

Mentzel's *Theatri Rerum Naturalium Brasiliae* vol. 4 also contains several pages with botanically unidentified oil paintings, which do not contain any written text, except the word "Anonyma" or a local name. As Marcgraf had already died and Mentzel was not a botanist, the latter probably did not know where to include these unnamed drawings. To the disappointment of twentieth-century scholars Whitehead and Boeseman, Mentzel did not indicate the name(s) of the person(s) that made these illustrations, but they assume that the artist(s) must have worked closely with Marcgraf.[78] In a letter, Johan Maurits claimed to have six painters in Brazil, but according to Brienen the oil paintings of the *Libri Picturati* were made by either Marcgraf himself or Albert Eckhout (c. 1607-c. 1666), painter of Brazilian still lifes and portraits of inhabitants of Dutch Brazil.[79] After studying the

Figure 4.5 (A) Blank page of Mentzel's *Theatri Rerum Naturalium Brasiliae*, vol. 4, reserved for copaiba and coapoiba (*Libri Picturati* A. 35: f. 77). Jagiellonian Library. (B) Detail of this page.

digital images of these unidentified paintings, we discovered that two of them are probably *Copaifera* species (Figures 4.6A and 4.7A).

The fruiting branch depicted in Figure 4.6A is unmistakably a *Copaifera* species, with the laterally compressed pods and the compound leaves, although they are imparipinnate, while *Copaifera* leaves are paripinnate. The number of leaflets in the painting is rather small, but leaves tend to drop off from dried specimens, as can be seen in the herbarium voucher in Figure 4.6B.

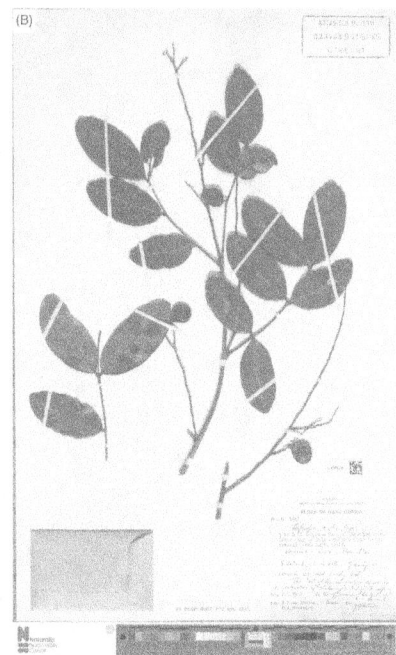

Figure 4.6 (A) Unidentified oil painting in Mentzel's *Theatri Rerum Naturalium Brasiliae*, vol. 4, showing resemblance to *Copaifera martii* (*Libri Picturati A 35: f. 231*). Jagiellonian Library. (B) Herbarium voucher of *C. martii* Hayne from Mato Grosso. Naturalis Biodiversity Center (U.1300158).

In the first detailed revision of Neotropical *Copaifera* species, Dwyer suggests that the species described by Marcgraf is *C. martii* Hayne, "especially as the leaflets are 'obrotunda aut etiam ovalia quattuor aut quinque digitos longa' (about 8–9 cm long), 'duos aut duos et semis lata' (about 3.6–5 cm wide); other characters suggestive of *C. martii* are the yellow arillus of the seed and the red bark."[80] In dry areas, *C. martii* takes the form of a shrub, as is mentioned on the label of Figure 4.6B, but in the forest it can grow as a tree up to 40 m high.[81] The species is widely distributed in Maranhão and northeastern Brazil.[82] In contrast, *C. officinalis* has more and larger leaflets, a white aril, and occurs mostly in the northern and central Amazon, Venezuela, and Colombia.[83]

The label on the specimen of *C. martii* depicted in Figure 4.6B indicates the local name of this species as "pau d'oi." The collector thinks that this vernacular name is a contraction of *olho* (eye) and probably refers to the appearance of the seed with its aril. It is more likely, however, that the local name is misspelled: pau-de-óleo is a common Brazilian name for *Copaifera* species.[84] Laboratory research has indicated that copaiba oil obtained from *C. martii*, collected in the state of Acre, exhibited good antibacterial activity against Gram-positive bacteria, including MRSA.[85]

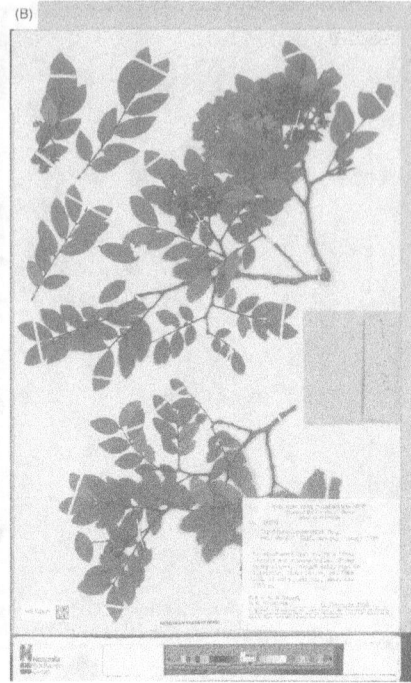

Figure 4.7 (A) Unidentified oil painting in Mentzel's *Theatri Rerum Naturalium Brasiliae*, vol. 4, showing resemblance to *Copaifera langsdorfii* Desf. (*Libri Picturati* A 35: f. 353). Jagiellonian Library. (B) Herbarium collection of *C. langsdorfii* from Minas Gerais. Naturalis Biodiversity Center (WAG.1639777).

The sterile branch depicted in Figure 4.7A is likely a member of the Leguminosae family, as has been written in pencil on the drawing by an unknown botanist, but due to the absence of fruits it is difficult to prove that it is a *Copaifera*. However, the leaves have long petioles, are sometimes paripinnate and have alternate leaflets with an obtuse apex. *Copaifera langsdorfii* can have up to 6 pairs of alternate or subopposite leaflets, petioles up to 9 cm, leaflets of up to 8 × 4 cm.[86] The brownish-green fruits are produced in large quantities and have one black seed with a yellow aril. The brown-red wood is used for construction.[87] The tree occurs from the Amazon to São Paulo, in different vegetation types, but is most commonly found in northeast Brazil.[88]

Initially, *C. officinalis* was thought to be the only species within the genus to produce the valuable oil.[89] Nowadays, more than 20 species of *Copaifera* yield copaiba oil in Brazil, but the most common supplier of the medicinal oil in Maranhão is probably *C. langsdorffii*.[90] The German botanist and explorer Carl Friedrich Philipp von Martius (1794–1868) was the first to give detailed descriptions and illustrations of *C. martii*

Figure 4.8 Illustration of *C. martii* and *C. langsdorfii* by Carl von Martius (1870, vol. XV, part II, fasc. 50: plate 63). Digitized by CRIA 2005.

and *C. langsdorfii* (Figure 4.8), based on extensive fieldwork and herbarium vouchers, collected during his travels in the Amazon and northeast Brazil.[91]

The Fig Tree "Quapoiba"

In his entry on "coapoiba" or "pao gamelo," Marcgraf mentioned that this name referred to several species.[92] After his description of *Clusia nemorosa*, he mentioned "another species," that went under these

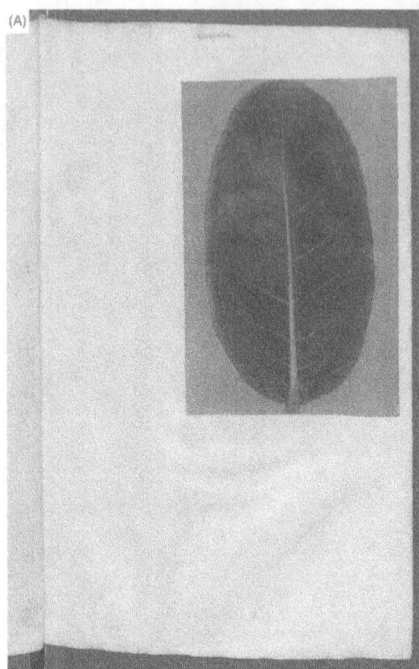
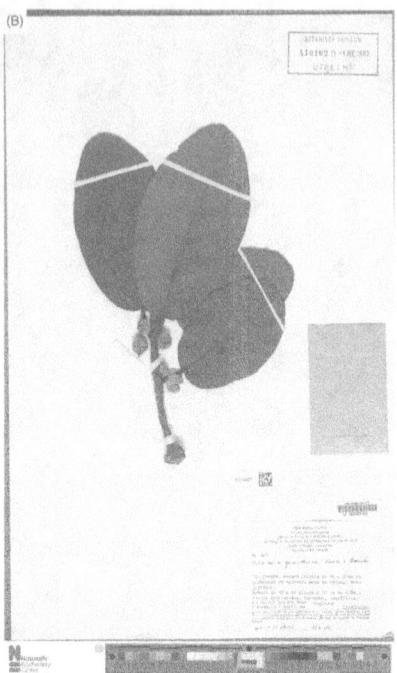

Figure 4.9 (A) Unidentified oil painting in Mentzel's *Theatri Rerum Naturalium Brasiliae,* vol. 4, showing resemblance to *Ficus gomelleira* (*Libri Picturati* A 35: f. 157). Jagiellonian Library. B) Specimen of *F. gomelleira* collected in the Brazilian Amazon. Naturalis Biodiversity Center (U.1425821).

names, which was a large tree with gray bark and wide branches, and leaves that were oblong, keeled, and glabrous. Its fruits were the size of small balls, full of tiny grains, like a fig: dry and tasteless. They were eaten by some people, although not much appreciated. This species was identified by Pickel as *Ficus doliaria* (Miq.) Mart.,[93] which is now a synonym of *Ficus gomelleira* Kunth & C.D. Bouché, a tree that is still known in Brazil as gameleira branca,[94] but also as copaibuçu or copaíba grande.[95]

In the collection of oil paintings in the *Libri Picturati*, there is also an unidentified illustration of a single leaf, with the local name "Quapoiba" written on it (Figure 4.9A), which bears a close resemblance to *Ficus gomelleira* (Figure 4.9B).

The fact that both *Clusia nemorosa* and *Ficus gomelleira* have large leathery leaves and sticky white exudate may have led to their shared local names, although the fruits and flowers of the two species are clearly different. According to Veiga Junior and Pinto, *F. gomelleira* has a similar crown as *Copaifera martii* when growing in open areas and therefore is named copaíba grande.[96] This confusing allocation of the names copaíba, coapoiba and gameleira also explains why around 1600 the Jesuit priest Cardim made a reference to a fig tree when he described the oil-producing "cupaigba" (Table 4.1), which was later copied by Manoel Tristão and De Laet.[97]

There is a small pencil note written on the drawing that says "Copiiba, Marcg. 121," which refers to the description of copiiba (*Tapirira guianensis* Aubl.) by Marcgraf.[98] Apart from the similar local name, the two species are unrelated and do not look alike, as *T. guianensis* has compound leaves with small leaflets and small, edible black fruits.

Conclusion

While the HNB may be the earliest published account of the Brazilian flora and fauna written by what could presently be understood as "trained scientists,"[99] our review indicates that the HNB was certainly not the first to report on medicinal plants from that area. The pioneering work of the Portuguese Jesuits remained largely unpublished, while the achievement of the HNB surpassed the Portuguese manuscripts with regard to detail, clarity, and scientific method.[100] Given the existing early reports on valuable natural resources that could be obtained from Brazil, Marcgraf and Piso probably had a wish list of useful plants to search for in the surroundings of Recife, which they were expected to describe and depict in more scientific detail. Partly due to the financial problems of Johan Maurits and his entourage, the large collection of natural history objects, descriptions, and illustrations produced in Dutch Brazil was later scattered across Europe.[101] Marcgraf's early death also hindered the botanical verification of unannotated botanical illustrations and their association with the Latin descriptions.

Some decades ago, Whitehead already suggested that the botanical study of the rest of the *Libri Picturati* would facilitate the identification of the flora and fauna described in the HNB.[102] The recent digitization of these paintings will make this feasible without having to examine the physical collections for a prolonged period. Plant species described in the HNB that currently lack taxonomic names can probably be identified by using these illustrations.

The HNB has long served as a naturalist's vade mecum for Brazil and other Neotropical regions. The Dutch, expelled from Brazil in 1654, started to explore the riches in the Guianas, using the HNB as a handbook to identify useful plants, as can be seen from the Brazilian local names used in their reports.[103] From their fortified trading posts, the Dutch exchanged knives, beads, alcohol, and plant products such as copaiba balsam with local Indigenous peoples.[104] The first botanist in Suriname, the enigmatic Hendrik Meyer, tried to collect a specimen of "Copayva" but did not know how the tree looked, so he ended up with a branch of *Neea constricta* Spruce ex J.A. Schmidt instead.[105]

The lack of clear descriptions on the botanical origin of copaiba balsam certainly did not hinder the trade in this precious medicinal product. From at least the seventeenth century, the balsam was widely exported to Portuguese trading posts in Asia (Goa and Macau), North America, and Europe, where it was sold in pharmacies.[106] A total of 21 pounds of "bals: copaiv." was transported on the Dutch ship *Wolphaartsdijk* from Cape of Good Hope to Batavia (Jakarta, Indonesia) and arrived there on 11 January 1729.[107] In his manuscript on the materia medica traded in Amsterdam around 1800, an anonymous Dutch merchant wrote that "Copaivae-Balsamum (*Copaifera officinalis*) comes from Brazil, from a tree growing on the island Maranhon and on the Antillean islands, which they call Copaiva-tree, from which this balsam flows. When it is of good quality it should dissolve entirely in Tri-Tartar."[108] This trade information indicates that copaiba balsam in Europe came from various sources and was probably of mixed origin at the time it entered the pharmacies.

Nowadays, Brazil remains the main exporter of copaiba balsam in the world, and the oil is used industrially for soap, lacquer, varnish, natural fragrances, and perfume.[109] Modern pharmacological studies have shown that it has anti-inflammatory, antibacterial, and analgesic properties that make scars heal faster and repel insects.[110] Many of these studies, however, have been conducted with commercially available copaiba oil, of which the botanical origin was unclear. Significant differences exist in the chemical composition of the more than 20 species of *Copaifera* that are tapped for their medicinal oil and even between individuals of the same species.[111] Further research is needed into the differences in composition and pharmacological properties of the oil of the *Copaifera* species in various Brazilian regions, but detailed descriptions and botanical illustrations are not yet available for every species. The work of Marcgraf and Piso is therefore far from finished.

Acknowledgment

We are grateful to Izabela Korczyńska, curator of the Jagiellonian Library, Kraków, for giving us access to the digitized images of the *Libri Picturati*.

Notes

1   The research for this chapter is part of the ERC project *BRASILIAE. Indigenous Knowledge in the Making of Science*, directed by Dr. Mariana Françozo at Leiden University and funded by the European Research Council Horizon 2020 Research and Innovation Programme (Agreement No. 715423). Unless otherwise indicated, all translations in this chapter are our own.

2   Peter J.P. Whitehead, "The Original Drawings for the *Historia Naturalis Brasiliae* of Piso and Marcgrave (1648)," *Journal of the Society for the Bibliography of Natural History* 7, no. 4 (1976): 409–422, doi: 10.3366/jsbnh.1976.7.4.409.

3 Rebecca Parker Brienen, *Visions of Savage Paradise: Albert Eckhout, Court Painter in Colonial Dutch Brazil* (Amsterdam: Amsterdam University Press, 2006); Harold J. Cook, *Matters of Exchange: Commerce, Medicine, and Science in the Dutch Golden Age* (New Haven, CT: Yale University Press, 2007).

4 Mariana Françozo, "Alguns Comentários à *Historia Naturalis Brasiliae*," *Cadernos de Etnolingüística* 2, no. 1 (2010): 1–7; Timothy D. Walker, "The Medicines Trade in the Portuguese Atlantic World: Acquisition and Dissemination of Healing Knowledge from Brazil (c. 1580–1800)," *Social History of Medicine* 26, no. 3 (2013): 403–431, doi: 10.1093/shm/hkt010.

5 Whitehead, "The Original Drawings."

6 Mireia Alcantara-Rodriguez, Mariana Françozo, and Tinde van Andel, "Plant Knowledge in the *Historia Naturalis Brasiliae* (1648): Retentions of Seventeenth-Century Plant Use in Brazil," *Economic Botany* 73, no. 3 (2019): 390–404, doi: 10.1007/s12231-019-09469-w.

7 Carlos Ossenbach, "Precursors of the Botanical Exploration of South America: Wilhelm Piso (1611–1678) and Georg Marcgrave (1610–1644)," *Lankesteriana* 17, no. 1 (2017): 61–71, doi: 10.15517/lank.v17i1.28527; Stephen A. Harris, "Snapshots of Tropical Diversity: Collecting Plants in Colonial and Imperial Brazil," in *Naturalists in the Field: Collecting, Recording and Preserving the Natural World from the Fifteenth to the Twenty-First Century*, ed. Arthur MacGregor (Leiden: Brill, 2018), 550–577.

8 Maarten Christenhusz, "The Hortus Siccus (1566) of Petrus Cadé: A Description of the Oldest Known Collection of Dried Plants Made in the Low Countries," *Archives of Natural History* 31, no. 1 (2004): 30–43, doi: 10.3366/anh.2004.31.1.30; Anastasia Stefanaki, Henk Porck, Ilaria Maria Grimaldi, Nikolaus Thurn, Valentina Pugliano, Adriaan Kardinaal, Jochem Salemink, et al., "Breaking the Silence of the 500-year-old Smiling Garden of Everlasting Flowers: The En Tibi Book Herbarium," *PLOS One* 14, no. 6 (2019): e0217779, doi: 10.1371/journal.pone.0217779.

9 Willem Piso and Georg Marcgraf, *Historia Naturalis Brasiliae: In qua non tantum Plantae et Animalia, sed et Indigenarum Morbi, Ingenia et Mores Describuntur et Iconibus supra Quingentas Illustrantur* (Leiden and Amsterdam: Elzevier, 1648); Willem Piso, *De India utriusque Re Naturali et Medica* (Amsterdam: Elzevier, 1658); Alcantara-Rodriguez et al., "Plant Knowledge;" see also Françozo in this volume.

10 Walker, "The Medicines Trade."

11 Stephen Snelders, *Vrijbuiters van de Heelkunde* (Amsterdam: Atlas Contact, 2012); Katrina Maydom, "New World Drugs in England's Early Empire" (PhD diss., University of Cambridge, 2019).

12 Cook, *Matters of Exchange*; Snelders, *Vrijbuiters van de Heelkunde*.

13 Eugene Gudger, "George Marcgrave, the First Student of American Natural History," *Popular Science Monthly* 81 (1912): 250–274.

14 Carl Linnaeus, *Systema Naturae*, vol. 1 (Stockholm: Laurentius Salvius, 1758); Whitehead, "The Original Drawings."

15 See Walker in this volume.

16 Walker, "The Medicines Trade;" Harris, "Snapshots of Tropical Diversity."

17 John Lloyd, "*Copaifera officinalis*: Botanical Description and Historical Notes," *Western Druggist* 20 (1898): 54–57; Tinde van Andel and Mireia Alcantara Rodriguez, "Vreemde Planten uit Verre Landen: Het Geheim van de Copaiba Olie uit Nederlands-Brazilië (1648)," in *Botanische Meesterwerken*, ed. Eddy Weeda, Joop Schaminee, and Nils van Rooijen (Zeist: KNNV, 2016), 133–144.

18   Bento José Pickel, *Flora do Nordeste do Brasil Segundo Piso e Marcgrave no Século XVII*, ed. Argus V. de Almeida (Recife: Universidade Federal Rural de Pernambuco, 2008); Alcantara-Rodriguez et al., "Plant Knowledge."

19   Whitehead, "The Original Drawings;" Jan de Koning, Gerda van Uffelen, Alicja Zemanek, and Bogdan Zemanek, *Drawn After Nature: The Complete Botanical Watercolours of the 16th-Century Libri Picturati* (Leiden: Brill, 2008).

20   Piso and Marcgraf, *Historia Naturalis*; Piso, *De India utriusque*; Mireia Alcantara Rodriguez, "Medicinal and Other Useful Plants from *Historia Naturalis Brasiliae* (1648): Are They Currently Used in Brazil?" (MSc thesis, Utrecht University, 2015); Alcantara-Rodriguez et al., "Plant Knowledge."

21   Piso and Marcgraf, *Historia Naturalis*; Piso, *De India utriusque*; George Marcgraf, *História Natural do Brasil* (São Paulo: Imprensa Oficial, 1942); Willem Piso, *História Natural do Brasil Ilustrada* (São Paulo: Companhia Editora Nacional, 1948); Willem Piso, *História Natural e Médica da Índia Ocidental* (Rio de Janeiro: I.N. Livro, 1957).

22   Dardano de Andrade Lima, Anne Fox Maule, Troels Pedersen, and Knut Rahn, "Marcgrave's Brazilian Herbarium, Collected 1638–44," *Botanisk Tidsskrift* 71 (1977): 121–160.

23   Pickel, *Flora do Nordeste*.

24   "Reflora," *JBRJ*, accessed 24 May 2022, http://floradobrasil.jbrj.gov.br.

25   "speciesLink," *CRIA*, accessed 24 May 2022, http://www.splink.org.br.

26   "Tropicos," *Missouri Botanical Garden*, accessed 24 May 2022, http://www.tropicos.org.

27   "BioPortal," *Naturalis Biodiversity Center*, accessed 24 May 2022, https://bioportal.naturalis.nl.

28   "GBIF," *Global Biodiversity Information Facility*, accessed 24 May 2022, https://www.gbif.org/en/.

29   "The Plant List (TPL)," *The Plant List*, accessed 24 May 2022, http://www.theplantlist.org/.

30   Van Andel and Alcantara Rodriguez, "Vreemde Planten."

31   Antônio G. da Cunha, *Dicionário Histórico das Palavras Portuguesas de Origem Tupi* (São Paulo: Edições Melhoramentos/Editora da Universidade de São Paulo, 1978).

32   Lloyd, "*Copaifera officinalis*;" Tinde van Andel, Sarina Veldman, Paul Maas, Gerard Thijsse, and Marcel Eurlings, "The Forgotten Hermann Herbarium: A 17th Century Collection of Useful Plants from Suriname," *Taxon* 61, no. 6 (2012): 1296–1304, doi: 10.1002/tax.616010.

33   "Time Capsule," *Utrecht University*, accessed 24 May 2022, http://timecapsule.science.uu.nl.

34   Whitehead, "The Original Drawings;" British Library, Manuscript Collection, "Sloane 1554Chartaceus, in folio, ff.81. sec XVII.Excerpta…" (Sloane Ms. 1554).

35   Lloyd, "*Copaifera officinalis*;" Otto Hartig, "Peter Martyr d'Anghiera," *The Catholic Encyclopedia* vol. IX (New York: Robert Appleton Company, 1910); Theodore Maynard, "Peter Martyr D'Anghiera: Humanist and Historian," *The Catholic Historical Review* 16, no. 4 (1931): 435–448.

36   Gabriel Soares de Sousa, *Notícia do Brasil: Comentários e Notas de F. A. de Varnhagen, Manuel A. Pirajá da Silva e Frederico Edelweiss* (São Paulo: Ed. Patrocinada pelo Departamento de Assuntos Culturais do M.E.C., 1974); Frederico Carlos Hoehne, *Botânica e Agricultura no Brasil no Século XVI* (São Paulo: Editora Nacional, 1937), 100; Serafim Leite, *Monumenta Brasiliae III*, serie Monumenta Missionum Societatis Iesu, vol. 81 (Rome: Institutum Historicum Societatis Iesu, 1958).

37 Works consulted were (in the order of authors' appearances on Table 4.1): Lloyd, "*Copaifera officinalis;*" Leite, *Monumenta Brasiliae*, 252; Pêro de Magalhães Gândavo, *História da Província Santa Cruz a que Vulgarmente Chamamos Brasil* (Lisboa: Antonio Gonsalves, 1576), 19; Jean de Léry, *Histoire d'un Voyage Faict en la Terre du Bresil, Autrement Dite Amerique* (Geneva: Antoine Chuppin, 1578), 179–180; Walker, "The Medicines Trade:" Soares de Sousa, *Notícia do Brasil*, 183–184, 325–326; Carolus Clusius, *Exoticorum Libri Decem, quibus Animalium, Plantarum, Aromatum, aliorumque Reregrinorum Fructuum Historiae Describuntur* (Leiden: Officina Plantiniana Raphelengii, 1605), 297; Fernão Cardim, *Tratados da Terra e Gente do Brasil* (Rio de Janeiro: J. Leite & Cia, 1925 [1548?-1625]), 62; Francisco Soares, "De Algumas Coisas Mais Notáveis do Brasil," *Revista do Instituto Histórico e Geographico Brasileiro* 94, vol. 148 (1923 [1594]): 409–410; Simão Travaços, "Declaração do Brasil: Livro Primeiro em que se Declara Toda a Costa, e Povoações do Estado do Brazil," (1596), manuscript in the library of Dr. J.J. Renoux, Rio de Janeiro, cited in Cunha, *Dicionário Histórico*; Ambrósio Fernandes Brandão, *Dialogues of the Great Things of Brazil*, trans. and annotated by Frederick H. Hall, William F. Harrison, and Dorothy Winters Welker (Albuquerque: University of New Mexico Press, 1987 [c. 1555]); Samuel Purchas, *Purchas His Pilgrimes*, vol. II (London, UK: William Stansby for Henrie Fetherstone, 1625), 1308; Johannes de Laet, *Nieuvve Wereldt ofte Beschrijvinghe van West-Indien* (Leiden: Elzevier, 1625), 208; Johannes de Laet, *Novus Orbis seu Descriptionis Indiae Occidentalis* (Leiden: Elzevier, 1633), 559–560; Johannes de Laet, *L'Histoire du Nouveau Monde ou Description des Indes Occidentales* (Leiden: Elzevier, 1640), 494; *Pharmacopoea Amstelredamensis, Senatus Auctoritate Munita, & Recognita. Editio Quarta* (Amsterdam: Iohannes Blaeu, 1643), 15; Caspar Barlaeus, *Rerum per Octennium in Brasilia [...]* (Amsterdam: Iohannes Blaeu, 1647), 134; Piso and Marcgraf, *Historia Naturalis*, part II, 130; Piso and Marcgraf, *Historia Naturalis*, part I, 23, 28, 30, 36, 39, 56–57; Piso, *De India utriusque*, 118.

38 Cardim, *Tratados da Terra*.
39 Purchas, *Purchas His Pilgrimes*, 1308; Lloyd, "*Copaifera officinalis.*"
40 Fernandes Brandão, *Dialogues*, 107–113; see Walker in this volume.
41 Soares de Sousa, *Notícia do Brasil*.
42 Soares, "De Algumas Coisas," 409–410.
43 Barlaeus, *Rerum per Octennium*, 226.
44 Lloyd, "*Copaifera officinalis.*"
45 *Pharmacopoea Amstelredamensis*, 15; *Pharmacopoea Leidensis, Amplissimorum Magistratuum Auctoritate Instaurata* (Leiden: Samuel Luchtmans, 1718).
46 De Laet, *Nieuvve Wereldt*; De Laet, *Novus Orbis*; De Laet, *L'Histoire*.
47 Clusius, *Exoticorum Libri Decem*; Cristóbal de Acosta, *Tractado de las Drogas y Medicinas de la Indias Orientales* (Burgos: Martin de Victoria, 1578).
48 John Carter Brown Library, Providence, Manuscript Collection, Codex Du 1, "*Beschrijvinge van de Custe van Brasil;*" Benjamin Teensma, *Suiker, Verfhout en Tabak: Het Braziliaanse Handboek van Johannes de Laet* (Zutphen, Walburg Pers, 2009).
49 Teensma, *Suiker, Verfhout en Tabak*, 73.
50 Alcantara-Rodriguez et al., "Plant Knowledge."
51 Piso and Marcgraf, *Historia Naturalis*, part I, 56–57.
52 Fernando São Paulo, "Comentários Sobre os Dois Primeiros Livros de Piso," in *História Natural do Brasil Ilustrada*, Willem Piso (São Paulo: Companhia Editora Nacional, 1948), 356–360.
53 Piso and Marcgraf, *Historia Naturalis*, part I, 23.
54 Ibid, 28, 30.

55   Ibid, 36.
56   Ibid, 57.
57   Ibid, 39.
58   Ibid, 57.
59   Maria Angélica da Silva and Melissa Mota Alcides, "Collecting and Framing the Wilderness: The Garden of Johan Maurits (1604–79) in North-East Brazil," *Garden History* 30 (2002): 153–176, doi: 10.2307/1587250.
60   Gudger, "George Marcgrave;" Whitehead, "The Original Drawings."
61   Peter Whitehead, "The Biography of Georg Marcgraf (1610–1643/4) by his Brother Christian, Translated by James Petiver," *Journal of the Society for the Bibliography of Natural History* 9 (1979): 301–314, doi: 10.3366/jsbnh.1979.9.3.301; Rebecca Parker Brienen, "Georg Marcgraf (1610-c.1644): A German Cartographer, Astronomer and Naturalist-Illustrator in Colonial Dutch Brazil," *Itinerario* 25, no.1 (2001): 85–122, doi: 10.1017/S0165115300005581; Cook, *Matters of Exchange.*
62   Brienen, "Georg Marcgraf;" Ossenbach, "Precursors."
63   British Library, "Sloane 1554Chartaceus," f. 68; Piso and Marcgraf, *Historia Naturalis*, part II, 130–131.
64   Cook, *Matters of Exchange.*
65   Lloyd, "*Copaifera officinalis*;" Pickel, *Flora do Nordeste*; Maria Franco Trindade Medeiros and Ulysses Paulino Albuquerque, "Food Flora in 17th Century: Northeast Region of Brazil in *Historia Naturalis Brasiliae*," *Journal of Ethnobiology and Ethnomedicine* 10 (2014): 50, doi: 10.1186/1746-4269-10-50; Alcantara-Rodriguez et al., "Plant Knowledge."
66   Piso and Marcgraf, *Historia Naturalis*, part I, 118.
67   Ibid, part II, 131.
68   Raphael Bluteau, *Vocabulario Portuguez e Latino* (Coimbra: Real Collegio das Artes da Companhia, 1713).
69   Valdir F. Veiga Junior and Angelo C. Pinto, "O Gênero *Copaifera* L.," *Química Nova* 25, no. 2 (2002): 273–286, doi: 10.1590/S0100-40422002000200016.
70   Pickel, *Flora do Nordeste.*
71   Nicolai J. von Jacquin, *Enumeratio Systematica Plantarum, quas in Insulis Caribaeis Vicinaque Americes Continente Detexit Novas, aut iam Cofnitas Emendavit* (Leiden: Theodoor Haak, 1760), 4; Nicolai J. von Jacquin, *Selectarum Stirpium Americanarum Historia* (Vienna: Krausiana, 1763), 133–134; Lloyd, "*Copaifera officinalis.*"
72   Carl Linnaeus, *Species Plantarum*, ed. 2, vol. 1 (Laurentius Salvius: Stockholm, 1762), 557.
73   Fusée Aublet, *Histoire des Plantes de la Guiane Françoise* (Paris: P.F. Didot, 1775), 399.
74   Carl Linnaeus, *Systema, Genera, Species Plantarum*, vol. 1 (Leipzig: Otto Wigand, 1840), 414.
75   Manuel A. Pirajá da Silva, "Introdução, Comentários e Notas da Obra" in Soares de Sousa, *Notícia do Brasil.*
76   Peter J. P. Whitehead and Marinus Boeseman, *A Portrait of Dutch 17th Century Brazil: Animals, Plants and People by the Artists of Johan Maurits of Nassau* (Amsterdam: North Holland Publishing Company, 1989).
77   Whitehead, "The Original Drawings."
78   Whitehead and Boeseman, *Dutch 17th Century Brazil.*
79   Whitehead, "The Original Drawings;" Brienen, "Georg Marcgraf;" for the recent identification of another painter in Dutch Brazil, see also Michiel van Groesen, "Abraham Willaerts: Marine Painter in Dutch Brazil and the Atlantic World," *Oud Holland* 132, nos. 2–3 (2019): 65–78, doi: 10.1163/18750176-1320203002.

80 John D. Dwyer, "The Central American, West Indian, and South American species of *Copaifera* (Caesalpiniaceae)," *Brittonia* 7, no. 3 (1951), 143–172, doi: 10.2307/2804703, 144; Piso and Marcgraf, *Historia Naturalis*, part II, 130–131.

81 Regina Martins-da-Silva, Jorge Fontella Pereira, Haroldo Cavalcante de Lima, "O Gênero *Copaifera* (Leguminosae-Caesalpinioideae) na Amazônia Brasileira," *Rodriguésia* 59, no. 3 (2008): 455–476, doi: 10.1590/2175-7860200859304.

82 "*Copaifera martii* Hayne," *Global Biodiversity Information Facility*, accessed 24 May 2022, https://www.gbif.org/species/2978164.

83 "*Copaifera officinalis* L.," *Global Biodiversity Information Facility*, accessed 24 May 2022, https://www.gbif.org/en/species/2978155.

84 Fabio Alessandro Pieri, Maria Carolina Martins Mussi, Maria Aparecida Moreira, "Óleo de Copaíba (*Copaifera* sp.): Histórico, Extração, Aplicações Industriais e Propriedades Medicinais," *Revista Brasileira de Plantas Medicinais* 11, no. 4 (2009): 465–472, doi: 10.1590/S1516-05722009000400016.

85 Adriana Oliveira Santos, Tânia Ueda-Nakamura, Benedito Prado Dias Filho, Valdir F. Veiga Junior, Angelo C. Pinto, and Celso Vataru Nakamura, "Antimicrobial Activity of Brazilian Copaiba Oils Obtained from Different Species of the *Copaifera* Genus," *Memórias do Instituto Oswaldo Cruz* 103 (2008): 277–281, doi: 10.1590/S0074-02762008005000015.

86 Dwyer, "*Copaifera* (Caesalpiniaceae)," 143–172.

87 Harri Lorenzi, *Árvores Brasileiras: Manual de Indentificação e Cultivo de Plantas Arbóreas Nativas do Brasil* (Nova Odessa: Instituto Plantarum de Estudios da Flota LTDA, 2010).

88 Liana Geraldo Souza de Oliveira, Daiany Alves Ribeiro, Manuele Eufrasio Saraiva, Delmácia Gonçalves de Macêdo, Julimery Gonçalves Ferreira Macedo, Patricia Gonçalves Pinheiro, José Galberto Martins da Costa, Marta Mariade Almeida Souza, and Irwin RoseAlencar de Menezes, "Chemical Variability of Essential Oils of *Copaifera langsdorffii* Desf. in Different Phenological Phases on a Savannah in the Northeast, Ceará, Brazil," *Industrial Crops and Products* 97 (2017): 455–464, doi: 10.1016/j.indcrop.2016.12.031; "*Copaifera langsdorffii* Desf.," *Global Biodiversity Information Facility*, accessed 24 May 2022, https://www.gbif.org/species/2978171.

89 Lloyd, "*Copaifera officinalis.*"

90 Santos et al., "Antimicrobial Activity."

91 Carl von Martius, *Flora Brasiliensis*, vol. 15: II (London, UK: George Bethma, 1870), fasc. 50, plate 63.

92 Piso and Marcgraf, *Historia Naturalis*, part II, 132.

93 Pickel, *Flora do Nordeste*.

94 Manoel Pio Corrêa, *Dicionário de Plantas Uteis do Brasil e das Exóticas Cultivadas*, vol. III (Rio de Janeiro: Imprensa Nacional, 1978).

95 Veiga Junior and Pinto, "O Gênero *Copaifera.*"

96 Veiga Junior and Pinto, "O Gênero *Copaifera.*"

97 Purchas, *Purchas His Pilgrimes*; De Laet, *Nieuvve Wereldt*.

98 Piso and Marcgraf, *Historia Naturalis*, part II, 121.

99 Brienen, "Georg Marcgraf."

100 Walker, "The Medicines Trade."

101 Brienen, "Georg Marcgraf"; Mariana Françozo, *De Olinda a Holanda: O Gabinete de Curiosidades de Nassau* (Campinas: Ed. Unicamp, 2014).

102 Whitehead, "The Original Drawings."

103 Maria Sybilla Merian, *Metamorphosis Insectorum Surinamensium: Ofte Verandering der Surinaamsche Insecten: Waar in de Surinaamsche Rupsen en Wormen Met Alle des Zelfs Veranderingen Na het Leven Afgebeeld en Beschreven Worden Zynde Elk Geplaatst op die Gewassen, Bloemen en Vruchten Daar Zij op Gevonden Zijn* (Amsterdam: Gerard Valck, 1705).

104   James Rodway, *History of British Guiana, from the Year 1668 to the Present Time*, vol. 1 (Georgetown: J. Thomson, 1891); James Rodway, *History of British Guiana, from the Year 1782 to the Present Time*, vol. 2 (Georgetown: J. Thomson, 1893).

105   "Collectie / Collection 35–49," *Herbarium Paul Hermann 1646–1695*, accessed 24 May 2022, https://www.hermann-herbarium.nl/COLL2.html; Van Andel et al., "The Forgotten Hermann Herbarium."

106   Walker, "The Medicines Trade."

107   "Time Capsule: Cargo," *Utrecht University*, accessed 24 May 2022, http://timecapsule.science.uu.nl/timecapsule/#/cargo.

108   Ingeborg Swart, Mieke Beumer, Wouter Klein, and Tinde van Andel, "Bodies of the Plant and Animal Kingdom: An Illustrated Manuscript on Materia Medica in the Netherlands (ca. 1800)," *Journal of Ethnopharmacology* 237 (2019): 236–244, doi: 10.1016/j.jep.2019.03.051.

109   Sigmund Rehm and Gustav Espig, *Cultivated Plants of the Tropics and Sub-Tropics* (Weikersheim: Verlag Josef Margraf, 1991).

110   Benjamin Gilbert, Duclineia Teixeira, Eliane Carvalho, Eliete de Paula, Jislaine Pereira, José Luiz Ferreira, Maria Beatriz Almeida, Reinaldo Machado, and Vera Cascon, "Activities of the Pharmaceutical Technology Institute of the Oswaldo Cruz Foundation with Medicinal, Insecticidal and Insect Repellent Plants," *Anais da Academia Brasileira de Ciências* 71 (1999): 265–271; José Carlos Carvalho, Vera Cascon, Lucas Silva Possebon, Mariana S.S. Morimoto, Luiz Gustavo Cardoso, Maria Auxiliadora Coelho Kaplan, and Benjamin Gilbert, "Topical Anti-inflammatory and Analgesic Activities of *Copaifera duckei* Dwyer," *Phytotherapy Research* 19, no. 11 (2005): 946–950, doi: 10.1002/ptr.1762; Niele M. Gomes, Claudia M. Rezende, Silvia P. Fontes, Maria Eline Matheus and Patricia Fernandes, "Antinociceptive Activity of Amazonian Copaiba Oils," *Journal of Ethnopharmacology* 109 (2007): 486–492, doi: 10.1016/j.jep.2006.08.018.

111   Valdir Veiga Junior, Elaine C. Rosas, Maria Valéria Carvalho, Maria das Graças Henriques, and Angelo Pinto, "Chemical Composition and Anti-inflammatory Activity of Copaiba Oils from *Copaifera cearensis* Huber ex Ducke, *Copaifera reticulata* Ducke and *Copaifera multijuga* Hayne: A Comparative Study," *Journal of Ethnopharmacology* 112 (2007): 248–254, doi: 10.1016/j.jep.2007.03.005; Vera Cascon and Benjamin Gilbert, "Characterization of the Chemical Composition of Oleoresins of *Copaifera guianensis* Desf., *Copaifera duckei* Dwyer and *Copaifera multijuga* Hayne," *Phytochemistry* 55 (2000): 773–778, doi: 10.1016/S0031-9422(00)00284-3.

5   An Imaginary Brazilian Zoo

Traditions and Innovations in the Portrayal of Animals in the *Historia Naturalis Brasiliae*

Annemarieke Willemsen

He was passing the trunk of the mainmast, heading for the prow, when he saw the aviary. [...] An embarrassed Adam, he could give no names to these creatures, except the names of birds of his own hemisphere: that one is a heron, he said to himself, that a crane, a quail. But it was calling a goose a swan. Prelates with broad cardinal's trains and beaks shaped like alembics spread grass-coloured wings, swelling a rosy throat and revealing an azure breast, chanting in almost human sound.
— Umberto Eco, *The Island of the Day Before* (1995), 40–41

Introduction

In *The Island of the Day Before* by Umberto Eco, set in the summer of 1643, the main character Roberto, shipwrecked onto an abandoned Dutch ship in the Pacific Ocean, comes across a wealth of caged birds from the "New" World, secretly kept on board.[1] He finds he lacks the language to describe their shapes and colors, his only foothold being to compare them to the animals he knows from his own part of the world. We often imagine that European travellers to South America in the sixteenth and seventeenth centuries, confronted with the wealth of unknown animals and desiring to describe and depict them to "take them home" and "show them to the world," were as overwhelmed as Eco's Roberto. We think that they basically used their knowledge of familiar European animals, based on books more than on seeing them, to paint their picture. A close look at the animals in the *Historia Naturalis Brasiliae* (henceforth HNB), the famous 1648 printed and illustrated account of Brazil based on the government of Johan Maurits of Nassau-Siegen,[2] gives a different idea of how eyewitnesses described and depicted animals they had never seen before. The treatise itself can be better understood when it is considered to be as much rooted in tradition as it is modern.

The HNB is first and foremost about plants. They take up the full first half of the work. The depictions of people from Brazil, which have attracted a lot of scholarly attention, are a very small proportion. Much more room is devoted to animals than to humans, but these have only been

DOI: 10.4324/9781003362920-6

studied incidentally, and these studies were mostly focused on the authors of the images.[3] In this chapter, the animals in the HNB will be presented with three research questions in mind. First, what animals are included in the treatise, how are they depicted, and why are they there? Second, to what extent are the depictions and descriptions of these animals using European knowledge, based on a long tradition of *bestiaria*, and to what extent are they using Indigenous knowledge, gained in Brazil itself? And finally, what is the role of the animals in the HNB in the shaping of the European image of Brazilian animals? This chapter is largely based on autopsy of one black-and-white and one colored specimen of the HNB, both held in the Leiden University Libraries under shelfmarks THYSIA 2274 and 1407 B 3.

Animals of Eden

The HRNB[4] consists of eight books. The first three are on plants and trees, books four until seven on animals, and book eight covers country and inhabitants. The books on animals start on page 142 and end on page 259. They are devoted to, in order, book four *De Piscibus Brasiliae* (on Brazilian fishes), book five *De Avibus* (on birds), book six *De Quadrupedibus & Serpentibus* (on four-legged animals and snakes), and book seven *De Insectis* (on insects). According to the table of contents, book four on "fishes, either from the sea or from rivers, and also shells" contains no fewer than 106 depictions and 19 notes. Book five on birds has 54 images and eight notes, book six on quadrupeds and snakes holds 33 images and 18 notes, and book seven on insects has 29 images and eight notes. This means that there are 222 images of animals in the HRNB, half of them fish, with only 26 mammals. As a comparison, the three books on plants and trees hold 200 images (about the same amount as the animals, but spread over more pages) and book eight, about land and people, only five.

A first encounter with Brazilian animals is the title page of the HNB. In spite of the division within the treatise, this title page most prominently shows the Indigenous people of Brazil, as a seventeenth-century Adam and Eve, in a lush landscape with lots of animals, mostly mammals, but also birds, fish, and a snake. There are slight variations in the coloring of various specimens of the treatise; I describe the title page of copy 1407 B 3 (see Figure 5.1). Two monkeys hold the banner with the title, on the left a black one with a gray beard, on the right a light brown one with a black face. The snake curled around the left tree is yellow and black and there is a gray sloth in the next tree. A yellow-brown bird is flying in the middle, and in the trees on the right side are a red and blue parrot, a purple bird with yellow wings and a blue crown, and a bird with a blue head, a red body, and brown wings. In the lower left corner, under the man's feet, in the water flowing from a jug held by a river god (personifying the

Figure 5.1 Colored title page of *Historia Naturalis Brasiliae* (Piso and Marcgraf, 1648: fs). Leiden University Libraries (copy 1407 B 3).

Amazon), there is a gray hammerhead shark, a red crab, greenish, black, brown, and gray fishes, and an octopus in green and brown. In the lower right corner, under the woman's feet, there is a yellow and black turtle, a dark gray ant eater, and a light-gray mousey monkey with a black-and-white striped tail.

Many of these species can be identified, for instance the typical macaw, but the variations in the coloring underline that this is not necessarily the idea. As has been argued before, this title page is a deliberate "New World version" of the well-known images of Adam and Eve in paradise, and the vegetation and wildlife merely serve to evoke an image of a lush Amazonian Garden of Eden. When comparing the common black-and-white version of the treatise with the colored one, it can be said that the coloring makes the animals, and especially the people, more "flat" and obscures their details, and certainly does not make them more realistic. But the paradise-like nature of the image as a whole does work better in color.

Useful Poison and Edible Wildlife

After the title page, the first animals are encountered in part I, *De Medicina Brasiliensi* by the physician Willem Pies (1611–1678, known by his latinized name Piso), that begins the HNB. This part consists of four books: I *De Aëre, Aquis, & Locis* (on skies, waters, and places), II *De Morbis Endemiis* (on endemic diseases), III *De Venenatis & Antidotis* (on poisons and antidotes), and IV *De Facultatibus Simplicium* (on simple properties). The third book, which starts on page 39, holds images of five of the snakes (Boicinininga, Cucurucu, Boiguacu, Iararaca, and Iniboboca),[5] a centipede (Scolopendra),[6] a sea cucumber (Moucicu),[7] a toadfish (Niqui),[8] and a toad (Cururu).[9] These animals have been included and depicted in this medicine book because of their poison that can be used. This proved to be one of the most valued aspects of the work, quoted in the eighteenth century as a rare source of information on snake poisons. For instance, Thomas Dancer wrote while in Nicaragua: "Piso reckons about twenty different species in Brazil: which I should suppose are most of them also inhabitants of this part of the coast. This part of Natural History, though in the highest degree interesting to the human species, has not been sufficiently cultivated. We are still, in a great measure, unacquainted with these noxious animals; and it is an object that claims the attention of natural inquirers, to investigate more particularly the species and distinctions of these reptiles, together with the proper antidotes against their several poisons."[10]

Piso uses the Brazilian names for these animals, and only rarely knows a Portuguese name ("Lusitanis") for (roughly) the same animal. His presentation of these venomous animals in a way they could be recognized (and used) is in line with his text. The diseases and Indigenous remedies he encountered in Brazil are well described; "for its time, it is an excellent example of reporting from observation."[11] His selection of animals is based on their use and this text clearly does not intend to demonstrate the wealth and wonders of the "New" World but rather to be a practical guide.

In the HRNB, books four to seven on animals start with fishes[12] and crustaceans,[13] then birds,[14] then quadrupeds,[15] lizards[16] and snakes,[17] and finally insects.[18] The text is in Latin, written by Georg Marcgraf (1610–1643) and edited by Johannes de Laet (1582–1649), director of the West India Company. It is very detailed, factual, and systematic: "when Marcgraf described an animal or a plant he resisted the temptation to leap to the most obvious characters and instead followed a quite rigid plan that obliged him to consider in turn all the less striking features; in adopting this very modern approach, one can be fairly certain that Marcgraf was actually composing his Latin description with the animal or plant before him and was not introducing hearsay."[19]

The woodcuts in the treatise are based mainly on drawings, sketches, and watercolors made in Brazil by Marcgraf[20] and by the talented painters Albert Eckhout (c.1610–1664) and Frans Post (1612–1680), who were also members of the elaborate entourage of Count Johan Maurits and part of its scientific program; they were "scientists with brushes" and "truly expedition artists."[21] Their drawings, of which many are preserved,[22] are not the first depictions of a Brazilian animal ever made, but they are the first large set and the species are very carefully depicted.

Jaguar and Blue-and-Yellow Macaw

Two examples here serve to illustrate a typical entry in the HRNB and its depiction: the jaguar and the macaw, both iconic Brazilian animals. One of the most impressive animals of Brazil must have been the jaguar (*Panthera onca*), of the Felid family, the biggest cat in the Americas. It is described: "Iaguara Brasiliensis [of Brazil], our tiger, *onca* in Portuguese. This animal is of the size of a wolf; however, larger ones are found. The head is thick and similar to the cat; the beard as well as the eyelids are similar to that of felines; the ears are short and round, almost like the cat's; the legs also imitate those of the cat; the feet are wide and have five fingers similar to those of the cat."[23] It proceeds with describing these paws with the number of nails, and how they can be subtracted, followed by the cat's teeth and spotted skin. After this long technical description, only the five last lines are devoted to the jaguar's cruel nature and how it kills and eats men and animals, to end with the sound that jaguars use to call each other in the night: "u u u." The adjoining woodcut shows a spotted cat, but rather plump, with thick, short legs, and a very short tail. The snout is quite long and thin and the ears are pointy. It does not look very athletic. Of the jaguar, also a drawing made by Frans Post in Brazil has been preserved (Figures 5.2 and 5.3). This looks undeniably more like the real animal as we know it, with the slender body, short snout, and long tail. The drawing is inscribed with a short caption in Dutch that translates as: "a tiger, as large as a common calf, they are very ferocious, destructive and strong, of this species there are some that are black."

110   *Annemarieke Willemsen*

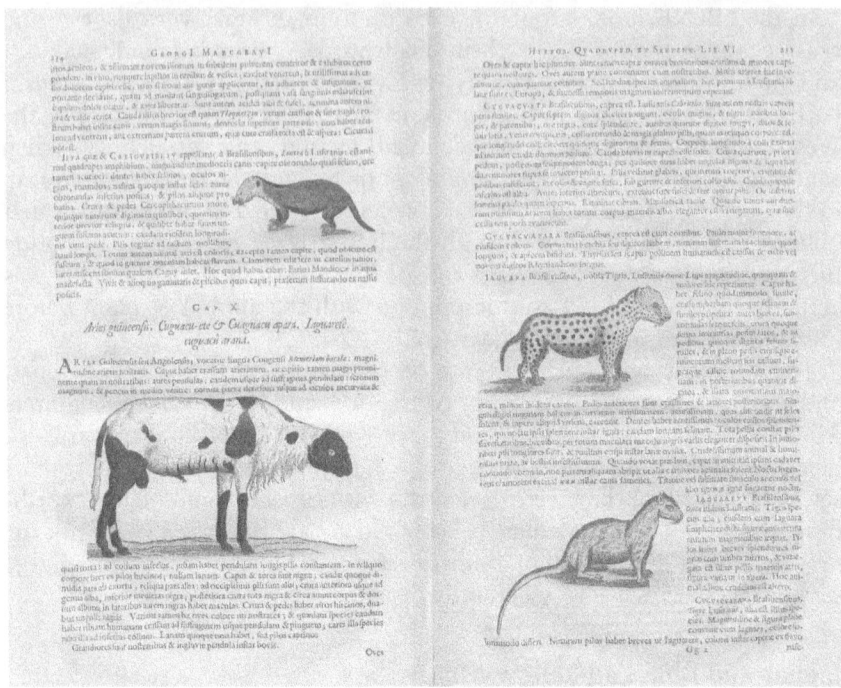

Figure 5.2 Mammals, including the jaguar, in *Historia Naturalis Brasiliae* (Piso and Marcgraf, 1648: part II, 234–235). Leiden University Libraries (copy 1407 B 3).

Figure 5.3 Jaguar, by Frans Post, watercolor and gouache, c.1638–1643. Noord-Hollands Archief/Beeldcollectie van de gemeente Haarlem (inv. no. 47058).

The blue-and-yellow macaw (*Ara ararauna*) of the Ara genus is called arara in Portuguese, after the Tupi name a'rara. This bird's description begins: "Ararauna Brasiliensis [of Brazil]. It is similar to the previous one in form but differs in color; the beak is black; the eyes, green-jay; the pupil is black. The white skin around the eyes is inlaid with black feathers, as if painted with a needle. The legs and feet are dusky; the head, on the front, above the beak has a tuft made of green feathers; under the lower part, black feathers surround the throat."[24] In this case, the text is purely descriptive of the bird and its feathers, and the lemmas on the other macaw species are focused on describing their plumage as well. The image shows a blue-and-yellow macaw standing on a wooden block; the bird is long and thin, with only one blue tail feather clearly shown. Macaws are also depicted on the title page of the HNB (see Figure 5.1).

Lifelike

In the prefaces, the illustrations are said to be "imagines ad vivum," so: drawn from life. Also, in Marcgraf's dedication to Johan Maurits, he states that he "accurately described the subjects of the book, with drawings made from life by himself."[25] What did this "from life" mean in the seventeenth century? Nowadays, we are used to extremely beautiful photos and films made in nature, with animals in their natural habitat. We are disappointed if an animal turns out to have been photographed or filmed in a zoo. But in the seventeenth-century animals were first and foremost shot to be studied and drawn. For exotic animals, this means that they were stuffed or preserved in alcohol and transported to Europe; we can safely assume that most zoologists of the seventeenth century never saw these animals move. They all used a long tradition of books on beasts to add knowledge and authority (and sometimes fiction) to any eyewitness accounts.

This does not mean that people before and in the seventeenth century were not able to see strange animals alive. Exotic animals, especially from Africa, had been part of the zoos in the gardens of royal and noble houses since the Early Middle Ages, and were exchanged more and more from the thirteenth century onwards.[26] We know elephants were paraded through the streets and displayed at fairs, while trainable animals like monkeys and parrots were kept by many rulers as part of their household. And again, there had been books about the "wonders of the world" all through the Middle Ages and their images were used widely. That made the public familiar with real and fabulous animals, including the likes of lions, tigers, elephants, and giraffes, as African and Asian animals had been known to Europeans for ages. This is why authors like Marcgraf could compare the features of his Brazilian animals to lions as easily as to pigs.

In Brazil, the shooting of species was sometimes done by the scientists or artists themselves, but mostly they relied on locals, who knew the animals in their habitat better and were also better at hunting them. Some of the drawings made by Frans Post in Brazil show dead animals, for instance an opossum and a rock cavy.[27] Next to hunting for science, artists were able to look at living animals in the zoo that was part of Johan Maurits' botanical gardens at Recife and is considered to have been the first zoo of the "New" World. Marcgraf will, for instance, have based many of his descriptions of the colors of animals on those kept in the zoo.[28] Johan Maurits also had a collection of stuffed animals they could look at, and some depictions of, for instance, birds on a small wooden pedestal seem to betray that they may have been drawn from stuffed specimens, like the "papagaij" (parrot) on page z16 of Handbook 1[29] and the blue-and-yellow macaw in the HNB (see Figure 5.4).

The zoo of Johan Maurits was described by Frei Manuel Calado, who writes that the Count "brought thither every kind of bird and animal that he could find, and since the local *moradores* [settlers] knew his taste and inclination, each one brought him whatever rare bird or beast he could find in the *sertão*, bringing him parrots, macaws, *jacús*, *cavindés*, *jaburus*, pheasants, Guinea fowl, ducks, swans, peacocks, turkeys, and great numbers of domestic fowls, and so many doves they could not be counted; there he had normal and black jaguars, pumas, ant-eaters, apes, *coatis*, squirrel monkeys, apereás, goats from Cape Verde, sheep from Angola, *cutias*, *pacas*, tapirs, wild boars, great numbers of rabbits, and in short there was nothing rare in Brazil that he did not have, since the *moradores* sent him these things willingly in view of his favour towards him."[30]

The drawings made in Brazil, or based on sketches made in the field, are in general of a better artistic quality than the later woodcuts. Within this corpus, the images of fishes and insects are generally more lifelike than those of birds and mammals. This is true not only for the woodcuts, but for the drawings and watercolors they were based on as well. This might be due to the circumstances these images were created in: insects were definitely caught, killed, and prepared to be depicted, as may have been the case with most of the fish.[31] Birds and mammals were probably more often portrayed from living or stuffed ones. It may also be the case that flatter, "more 2D" animals were just easier to draw than rounder, "more 3D" ones. It can be observed that the animals become "flatter" in the process: they are most lifelike in the sketches, and still very realistic in the watercolors; they lose volume when made into a woodcut, and even more when they are colored. While the uncharacteristically precise text of the HNB created by Marcgraf seems to have been reproduced by De Laet faithfully, the depictions made mostly by Post and Eckhout lost some (or sometimes much) of their lifelikeness in the editing process.

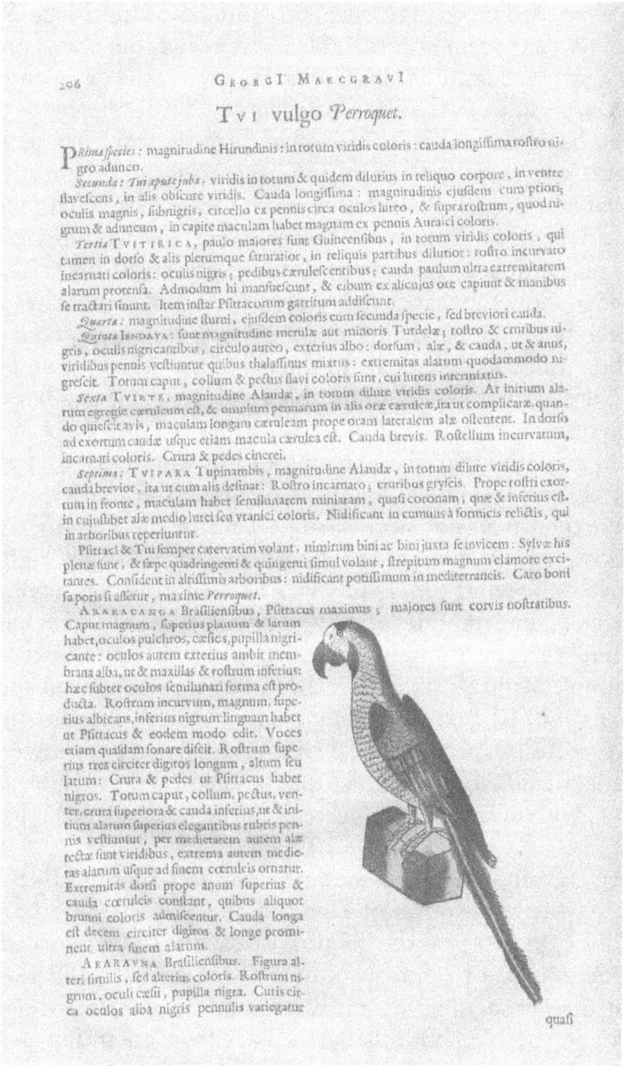
206 GEORGI MARCGRAVI

TVI vulgo *Perroquet.*

Prima species: magnitudine Hirundinis: in totum viridis coloris: cauda longissima rostro nigro adunco.

Secunda: Tui apute juba, viridis in totum & quidem dilutius in reliquo corpore, in ventre flavescens, in alis obscure viridis. Cauda longissima: magnitudinis ejusdem cum priori; oculis magnis, subnigris, circello ex pennis circa oculos luteo, & supra rostrum, quod nigrum & aduncum, in capite maculam habet magnam ex pennis Auraici coloris.

Tertia TVITIRICA, paulo majores sunt Guineensibus, in totum viridis coloris, qui tamen in dorso & alis plerumque saturatior, in reliquis partibus dilutior: rostro incurvato incarnati coloris: oculis nigris; pedibus cærulescentibus; cauda paulum ultra extremitatem alarum protensa. Admodum hi mansuescunt, & cibum ex alicujus ore capiunt & manibus se tractari sinunt. Item instar Psittacorum garritum addiscunt.

Quarta: magnitudine sturni, ejusdem coloris cum secunda specie, sed breviori cauda.

Quinta IENDAYA: sunt magnitudine merulæ aut minoris Turdelæ; rostro & cruribus nigris, oculis nigricantibus, circulo aureo, exterius albo: dorsum, alæ, & cauda, ut & anus, viridibus pennis vestiuntur quibus thalassinus mixtus: extremitas alarum quodammodo nigrescit. Totum caput, collum & pectus flavi coloris sunt, cui luteus intermixtus.

Sexta TVIETE, magnitudine Alaudæ, in totum dilute viridis coloris. At initium alarum egregie cæruleum est, & omnium pennarum in alis oræ cæruleæ, ita ut complicatæ, quando quiescit avis, maculam longam cæruleam prope oram lateralem alæ ostentent. In dorso ad exortum caudæ usque etiam macula cærulea est. Cauda brevis. Rostellum incurvatum, incarnati coloris. Crura & pedes cinerei.

Septima: TVIPARA Tupinambis, magnitudine Alaudæ, in totum dilute viridis coloris, cauda brevior, ita ut cum alis desinat: Rostro incarnato; cruribus gryseis. Prope rostri exortum in fronte, maculam habet semilunarem miniatam, quasi coronam; quæ & inferius est, in cujuslibet alæ medio lutei seu vranici coloris. Nidificant in cumulis à formicis relictis, qui in arboribus reperiuntur.

Psittaci & Tui semper catervatim volant, nimirum bini ac bini juxta se invicem: Sylvæ his plenæ sunt, & sæpe quadringenti & quingenti simul volant, strepitum magnum clamore excitantes. Consident in altissimis arboribus: nidificant potissimum in mediterraneis. Caro boni saporis si assetur, maxime *Perroquet.*

ARARACANGA Brasiliensibus, Psittacus maximus; majores sunt corvis nostratibus. Caput magnum, superius planum & latum habet, oculos pulchros, cæsios, pupilla nigricante: oculos autem exterius ambit membrana alba, ut & maxillas & rostrum inferius: hæc subter oculos semilunari forma est producta. Rostrum incurvum, magnum, superius albicans, inferius nigrum: linguam habet ut Psittacus & eodem modo edit. Voces etiam quasdam sonare discit. Rostrum superius tres circiter digitos longum, altum seu latum: Crura & pedes ut Psittacus habet nigros. Totum caput, collum, pectus, venter, crura superiora & cauda inferius, ut & initium alarum superius elegantibus rubris pennis vestiuntur, per medietatem autem alæ tectæ sunt viridibus, extrema autem medietas alarum usque ad finem cœruleis ornatur. Extremitas dorsi prope anum superius & cauda cœruleis constant, quibus aliquot brunni coloris admiscentur. Cauda longa est decem circiter digitos & longe prominent ultra finem alarum.

ARARAVNA Brasiliensibus. Figura alteri similis, sed alterius coloris. Rostrum nigrum, oculi cæsii, pupilla nigra. Cutis circa oculos alba nigris pennulis variegatur quasi

Figure 5.4 Blue-and-yellow macaw in *Historia Naturalis Brasiliae* (Piso and Marcgraf, 1648: part II, 206). Leiden University Libraries (copy 1407 B 3).

Animals Cut in Wood

While the title page of the HNB is a luxurious copper plate engraving, the hundreds of illustrations in the text are made with much cheaper and quicker woodcuts. They have been called "rather poor woodcuts" or "crudely drawn,"[32] but they are printed with clean blocks, very detailed, and sharp-lined. To our taste they may not represent the animals extremely realistically, but they are good illustrations. Also, the fact that a handful of woodcuts were used more than once, providing the same image for two different animals, may seem unforgivable for scientists but was quite common in printed encyclopedia, like the *Hortus Sanitatis*, the "garden of health," that was very popular in the sixteenth century.[33] The choice for these illustrations can be explained by the original purpose and intended use of the treatise and the expectations of the publisher as well as the reader. For like the medieval *bestiarium*, the HNB was not meant as a field guide, nor was it likely used as such. Almost no reader could and would compare these images to real animals, or even stuffed or preserved ones. These tomes were meant to show the reader the riches of creation (hence the appropriate paradise-like title page) and the reader expected to read and look and be amazed about this unknown part of the world and its exotic and wonderful animals. They did not expect to be able to identify a living beast from the description or depiction.

This also explains why the coloring of the woodcuts, and even of the title page, is inconsistent. It is meant to make the tome more beautiful (and more expensive), not necessarily to aid in the identification of the species portrayed. Of course, the colors had to be largely right but they did not have to be controllably precise. The coloring was always going to be subjective to interpretation, as it was not done by someone who had seen, or could see, the animals themself. They worked to instructions and maybe examples, and they did their best. But we know from specimens of the *Hortus Sanitatis* that were only partly colored, that the instructions given were quite unspecific, like "red" written next to flowers. That is why there is a range to how much of the bird furthest right on the title page is blue, with even a fully blue beak in the Leiden University Libraries copy.

A surprising glimpse into the way compilers of the HNB processed knowledge is their invention of Dutch names for the Brazilian animals, mentioned in the text of the lemma after the Indigenous and (where known) Portuguese or Spanish name. These animals did not yet have a name in the vernacular, which was made up for this occasion either in Brazil or in the Netherlands by De Laet from their appearance. This aspect of the Latin text has attracted little attention, as most researchers have been more interested in connecting proper Latin names to the species described in the HRNB. Good examples of these names are for instance the "cruyshaye," which is "kruishaai," translating as "cross shark."[34] There is a hummingbird that is given the quite poetic Dutch name "bloemen-specht," a "flower pecker," alluding to the woodpecker that pecks trees like a hummingbird pecks flowers.[35] Further, an armadillo, named "schild-vercken," or "shelled pig;"[36] this name is given in analogy to the Dutch word for turtle and tortoise, which is *schildpad*, the *schild-* meaning both shield and a shell that shields something, and the *-pad* being a toad. So, if the *schildpad* is a "shelled toad," so the armadillo is a "shelled pig." These names are based on the mental world of these seventeenth-century men, partly referring to animals they knew at home, like the woodpecker and tortoise, but evidently not based on their knowledge from older *bestiaria*. Rather, they refer to well-known daily items, the best example being the elongated, thin fish "Petimbuaba" that is called a "tabac-pijpe" or "tobacco pipe" because of its shape and color (see Figure 6.1).[37] It is still called a "pipe-fish."

Traditional or Modern

The HNB has been called the first natural history of the "New" World and Marcgraf is considered the first student of American natural history.[38] The treatise is seen as the beginning of a scientific tradition and as the first in a series of zoological publications. As we have seen, however, the tome was not meant for this and is only partly suited for it. Importantly, the publication also follows in a long tradition of books on the natural world, first hand-written and hand-painted, and afterwards printed in large quantities from shortly before 1500. The most printed and widely distributed European tome on plants and animals of the sixteenth century is the *Hortus Sanitatis* already mentioned, adapted from manuscript sources and first printed in Mainz in 1491, followed by many editions and translations, for instance in Middle-Dutch as *Den Groten Herbarius* in 1514. It is also illustrated with woodcuts (one with every lemma) and a number of the specimens have been colored with ink. A comparison with this *Hortus Sanitatis* is helpful to show the innovations of the HNB, but also the tradition in which it stood and the aims and expectations of both the printer and the intended audience (Figure 5.5).

First, the HNB starts every lemma with the Indigenous Brazilian name, followed by the Portuguese name (if known) and the Dutch name (sometimes invented). This is different from the medieval tradition, which was based on Latin names and always gives the Latin name first or in vernacular editions second after the translated name. In these encyclopedias there is often also an explanation of the name, its provenance, and meaning that can be quite extensive. This is missing from the HNB, certainly because the Brazilian names had no tradition of explanation that was known to the western visitors.

Second, in the text of the HNB, the emphasis is on the description of the animal, with precise measurements and a lot of counting (especially of nails). Less space is devoted to the properties or use of the animal, although at the end it is often mentioned whether the animal can be eaten or not. In earlier encyclopedias most of the lemma was taken up by these properties and uses. In the original lemmas by Marcgraf there seems to have been almost no attention to uses: these are mostly found in the short annotations ("Annotatio") by De Laet, which is also where we find references to "authoritas" on the natural world like Ulisse Aldrovandi (1522–1605). Again, this may be because the animals of the "New" World lacked a tradition of properties and uses ascribed to them, at least to Europeans.

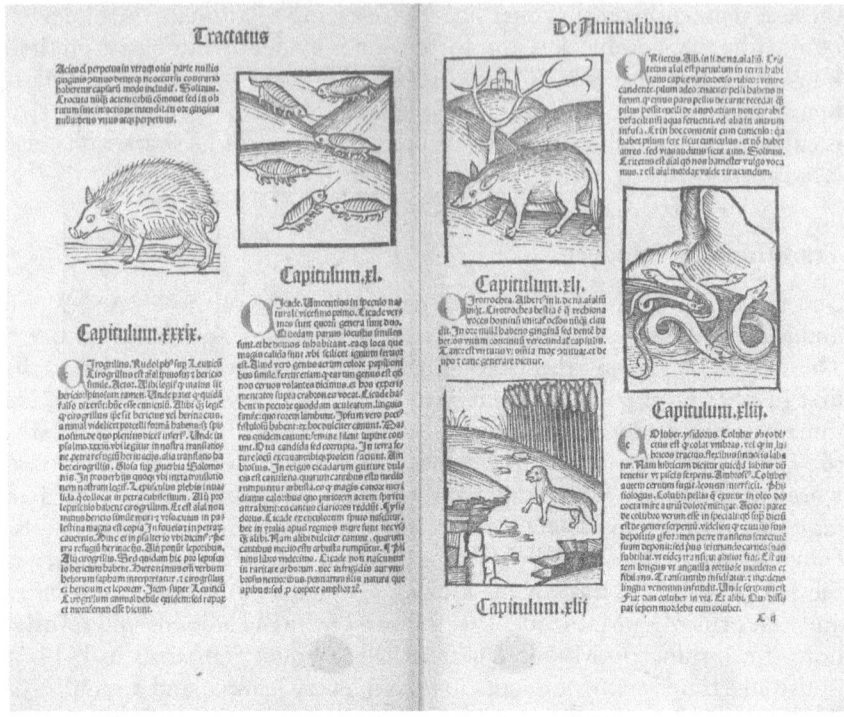

Figure 5.5 Various animals in the chapter "De Animalibus" of the *Hortus Sanitatis* (after 1491: 535–536). Smithsonian Libraries, Biodiversity Heritage Library (RS79. H82 1497; doi: 10.5962/bhl.title.61747).

Finally, completely missing from the HNB – also noted by Whitehead[39] – are fabulous animals or "monsters." In medieval encyclopedias, these were common and treated no different from other animals. In earlier written accounts on Brazil, and undoubtedly in stories told to the scientists, there were definitely monsters and magical animals. Marcgraf did not include these, simply because he worked from specimens he could see, rather than working from stories or in an established tradition like medieval compilers. Whitehead notes that the pace in which the drawings were worked into a publishable volume may have prevented compiler De Laet from including fabulous animals into the HNB at a later stage.

The emphasis on local names and precise descriptions of animals and the lack of "fabulous" aspects are the main reasons why the HNB was used by later scientists, most famously by Carl Linnaeus (1707–1778) in his classifications. These later scientists could use the detailed text for their work, but they had trouble determining the species from the name and description, and they complained (as many modern authors do) about the "poor illustrations:" "its animals and plants were frequently cited by Linnaeus and later authors, but identification is often hindered by the poor descriptions and woodcut illustrations."[40] This betrays that they were making "modern" use of a tome that was not meant to be scientific in their sense of the word. In the treatment of the illustrations, the HNB is well comparable to older *bestiaria*.

The Versatile Capybara

As no one had seen these animals depicted before and no one had access to the drawings or even the handful of depictions made before the publication of the HNB, its images, like its text, functioned as the best account of Brazilian animals for later science. This warrants the question what the impact of these images has been on the public image of the animals of Brazil. This is looked at through a case study featuring the capybara (*Hydrochoerus hydrochaeris*) of the *Caviidae* (guinea pig) family, the largest rodent in the world, also called "water hog." It is native to most of South America, but not to Central America, and was first seen by European explorers in Brazil (Figure 5.6).

The description in the HNB of this animal reads as follows: "Capy-bara Brasiliensis [of Brazil], river boar, similar in figure to the domestic hog, the size of one of our one- or two-year-old pigs. The length of the head to the anus is about two feet, the thickness of the belly one and a half feet. There is no tail. The feet are four and similar to those of the pig; the rear

Figure 5.6 Capybara in Poconé, Mato Grosso, Brazil, 2016. Photo by Bernard Dupont, "Capybara (Hydrochoerus hydrochaeris)," Flickr.

ones have four nails; the front ones, three; on the rear ones there is a very long middle nail, two shorter ones and a very small one; on the front legs the middle nail is longer and the other two, shorter; the skin is thick up until the curve of the leg. The length of the head is ten fingers and the same measure is valid for its thickness; this head is large and not proportioned; the mouth is very long and large; the eyes are large and black; the ears small and rounded. The lower jaw is shorter than the upper one; on both, there are two curved teeth, located one and a half fingers outside their cavity, and their length is of almost two fingers below the gum. These teeth, however, do not come out of the mouth but stay inward, as with hares. The other teeth are curious; in each jaw each set consists of eight bones, that is, four on each side; each bone represents three inseparable teeth. In this way the teeth are twenty-four in each jaw, making a total of forty-eight. They are all flat at the extremities. These animals eat grass and various fruits. Their meat can be eaten, but it is not tasty; roast is a little better, especially the head. They walk in great numbers along the banks of the S. Francisco River; they can swim and dive very well; at night they make a great awful cry, as donkeys usually do."[41]

Frans Post already painted a capybara prominently in the foreground of his view of the São Francisco river in 1639, a decade before the HNB was published (see Figure 5.7). He based this painting on drawings he made in Brazil, including a graphite sketch and a watercolor and gouache of the

Figure 5.7 View of the Rio São Francisco in Brazil with a capybara, by Frans Post, oil painting, 1639. Musée du Louvre (inv. no. 1727; B 300). Photo © RMN-Grand Palais (musée du Louvre)/René-Gabriel Ojeda.

rodent. These two drawings are part of a set of 34, only identified in 2015 in the provincial archives of Noord-Holland, that are considered the "missing link between Post's seven-year Brazilian adventure and the paintings he produced on his return to Haarlem. These drawings with their inscriptions have an immediacy about them that makes you feel as if you were looking over Frans Post's shoulder, as he recorded the fascinating fauna of the new world."[42] Compared to modern-day photographs of the animal (see Figure 5.6), the capybara on both the drawings and the painting is quite realistic, with the thick muzzle, the small rounded ears, the plump body, and the short and rather thin legs, all in a brown tint that varies over the body. Showing the animal at the edge of a river, as if in the middle of drinking water, is a realistic setting for this "river hog," as it was often called in this time. The description by Marcgraf also mentions specifically how lots of capybara walk the banks of this river. The capybara on the painting by Post is certainly more convincing than the woodcut in the HNB of 1648, which shows a seated animal with thick legs and an elongated narrow snout with two pointed ears (see Figure 5.8). The coloring does not help, as it is painted in one consistent, too dark reddish-brown color, with a white muzzle, while in reality the muzzle is darker than the face.

The painting is one of more than 100 Brazilian scenes painted by Frans Post after his return to the Netherlands in 1644,[43] using his sketches, drawings, and watercolors from his stay in Brazil. These also served as models for tapestries, for instance the well-known series *Les Anciennes*

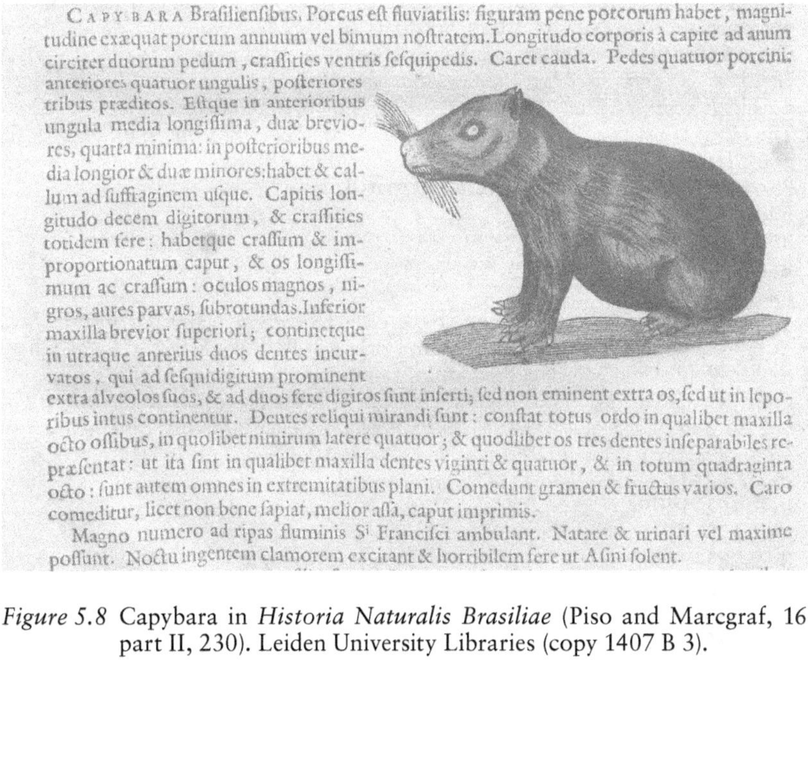
CAPYBARA Brasiliensibus. Porcus est fluviatilis: figuram pene porcorum habet, magnitudine exæquat porcum annuum vel bimum nostratem. Longitudo corporis à capite ad anum circiter duorum pedum, crassities ventris sesquipedis. Caret cauda. Pedes quatuor porcini: anteriores quatuor ungulis, posteriores tribus præditos. Estque in anterioribus ungula media longissima, duæ breviores, quarta minima: in posterioribus media longior & duæ minores: habet & callum ad suffraginem usque. Capitis longitudo decem digitorum, & crassities totidem fere: habetque crassum & improportionatum caput, & os longissimum ac crassum: oculos magnos, nigros, aures parvas, subrotundas. Inferior maxilla brevior superiori; continetque in utraque anterius duos dentes incurvatos, qui ad sesquidigitum prominent extra alveolos suos, & ad duos fere digitos sunt inserti; sed non eminent extra os, sed ut in leporibus intus continentur. Dentes reliqui mirandi sunt: constat totus ordo in qualibet maxilla octo ossibus, in quolibet nimirum latere quatuor; & quodlibet os tres dentes inseparabiles repræsentat: ut ita sint in qualibet maxilla dentes viginti & quatuor, & in totum quadraginta octo: sunt autem omnes in extremitatibus plani. Comedunt gramen & fructus varios. Caro comeditur, licet non bene sapiat, melior assa, caput imprimis.

Magno numero ad ripas fluminis S[i] Francisci ambulant. Natare & urinari vel maxime possunt. Noctu ingentem clamorem excitant & horribilem fere ut Asini solent.

Figure 5.8 Capybara in *Historia Naturalis Brasiliae* (Piso and Marcgraf, 1648: part II, 230). Leiden University Libraries (copy 1407 B 3).

Indes, of which a full set survives in the Palace of the Knights of Saint John in Valetta, Malta.[44] Some of the Brazilian paintings were shown in the Mauritshuis in The Hague in the seventeenth century, some were given to relations then, who displayed them as well, and nowadays they are spread over collections throughout the whole world.[45] It must be assumed that the paintings and tapestries reached a larger audience than the tomes, both in their own time and afterwards. The paintings by Post and Eckhout are still the best-known images from seventeenth-century Brazil. Therefore, while later scientists focused on the HNB as a first detailed and illustrated source for their classification of species, it was the monumental art based on the Brazilian campaign of Johan Maurits that shaped the common image of the natural world of South America. Ironically, that means that scientists have worked with less realistic depictions of the animals than other people will have seen.

Notes

1 Unless otherwise specified, all translations in this chapter are my own.
2 Willem Piso and Georg Marcgraf, *Historia Naturalis Brasiliae: In qua non tantum Plantae et Animalia, sed et Indigenarum Morbi, Ingenia et Mores Describuntur et Iconibus supra Quingentas Illustrantur* (Leiden and Amsterdam: Elzevier, 1648).
3 Ad. Schneider, "Die Vogelbilder zur Historia Naturalis Brasiliae des Georg Marcgrave," *Journal of Ornithology* 86 (1938): 74–106, doi: 10.1007/BF01982596; Rüdiger Joppien, "The Dutch Vision of Brazil: Johan Maurits and his Artists," in *Johan Maurits van Nassau-Siegen 1604–1679: A Humanist Prince in Europe and Brazil*, ed. Ernst van den Boogaart (The Hague: Johan Maurits van Nassau Stichting, 1979), 296–376; Peter J.P. Whitehead, "Georg Marcgraf and Brazilian Zoology" in Boogaart, *Johan Maurits*, 424–471; Marinus Boeseman, "A Hidden Early Source of Information on North-Eastern Brazilian Zoology," *Zoologische Mededelingen* 68 (1994): 113–125; Rebecca Parker Brienen, "Georg Marcgraf (1610–c.1644): A German Cartographer, Astronomer and Naturalist-Illustrator in Colonial Dutch Brazil," *Itinerario* 25, no. 1 (2001): 85–122, doi: 10.1017/S0165115300005581.
4 The HNB consists of two parts, I: Piso's *De Medicina Brasiliensi* and II: Marcgraf's *Historiae Rerum Naturalium Brasiliae* (HRNB).
5 Piso and Marcgraf, *Historia Naturalis*, part I, 41–42.
6 Ibid, 44.
7 Ibid.
8 Ibid, 45.
9 Ibid, 46.
10 Thomas Dancer, *A Brief History of the Late Expedition Against Fort San Juan* (Kingston: D. Douglass and W. Aikman, 1781), 13–14. I would like to thank Christopher Blakley and Mariana Françozo for drawing my attention to this source.
11 Peter J.P. Whitehead, "The Original Drawings for the *Historia Naturalis Brasiliae* of Piso and Marcgrave (1648)," *Journal of the Society for the Bibliography of Natural History* 7, no. 4 (1976): 409–422, doi: 10.3366/jsbnh.1976.7.4.409, 411.
12 Piso and Marcgraf, *Historia Naturalis*, part II, 142–181.
13 Ibid, 182–189.
14 Ibid, 190–219.
15 Ibid, 221–235.
16 Ibid, 236–239.
17 Ibid, 240–244.
18 Ibid, 245–259.

19 Peter J.P Whitehead and Marinus Boeseman, *A Portrait of Dutch 17th Century Brazil: Animals, Plants and People by the Artists of Johan Maurits of Nassau* (Amsterdam: North Holland Publishing Company, 1989), 200.

20 The authorship of the woodcuts and drawings has been debated for a long time by scholars, e.g., Brienen, "Georg Marcgraf," and Smith in this volume; to my chapter however, that is not essential.

21 Whitehead and Boeseman, *Dutch 17th Century Brazil*, 202–203.

22 Alexander de Bruin, *Frans Post: Animals in Brazil* (Amsterdam: Rijksmuseum, 2016).

23 Piso and Marcgraf, *Historia Naturalis*, part II, 235.

24 Ibid, 206–207.

25 Whitehead, "Georg Marcgraf," 435.

26 Raymond van Uytven, *De Papegaai van de Paus: Mens en Dier in de Middeleeuwen* (Leuven/Zwolle: Davidsfonds, 2003).

27 De Bruin, *Frans Post*, 46, 51.

28 Whitehead, "Georg Marcgraf," 429.

29 Whitehead and Boeseman, *Dutch 17th Century Brazil,* 244, plate 13d.

30 Whitehead, "Georg Marcgraf," 429.

31 See also Smith in this volume.

32 Whitehead and Boeseman, *Dutch 17th Century Brazil,* 21; Whitehead, "The Original Drawings," 410.

33 *Hortus Sanitatis* (Mainz: Jacob Meydenbach, 1491); see "Hortus Sanitatis," *Wikipedia*, accessed 25 May 2022, https://en.wikipedia.org/wiki/Hortus_Sanitatis, with links to many digitized specimen online.

34 Piso and Marcgraf, *Historia Naturalis*, part II, 181.

35 Ibid, 196.

36 Ibid, 231.

37 Ibid, 148; see also Smith in this volume.

38 Whitehead, "Georg Marcgraf," 424.

39 Whitehead, "Georg Marcgraf," 442.

40 Whitehead, "The Original Drawings," 409.

41 Georg Marcgraf, *História Natural do Brasil* (São Paulo: Imprensa Oficial, 1942 [1648]), 230. Excerpt translated into English by Mariana Françozo.

42 Alexander de Bruin, quoted in Rijksmuseum, "Spectacular Discovery of Drawings by Frans Post: For Centuries Unrecognized," press release, 8 September 2016 [no longer accessible online].

43 Whitehead, "The Original Drawings," 411.

44 Whitehead, "The Original Drawings," 416; Carrie Anderson, "The Old Indies at the French Court: Johan Maurits' Gift to Louis XIV," *Early Modern Low Countries Journal* 3, no. 1 (2019): 32–59.

45 Mariana Françozo, "Global Connections: Johan Maurits of Nassau-Siegen's Collection of Curiosities," in *The Legacy of Dutch Brazil*, ed. Michiel van Groesen (Cambridge, UK: Cambridge University Press, 2014), 105–123.

6 Marcgraf's Fish in the *Historia Naturalis Brasiliae* and the Rhetorics of Autoptic Testimony[1]

Paul J. Smith

Of all those who described the natural history of distant lands in the sixteenth and seventeenth centuries, [Marcgraf] was assuredly the most intelligent and the most exact, and the one who most contributed to the natural history of fishes.[2]
— Georges Cuvier, *Historical Portrait of the Progress of Ichthyology*
(1995 [1828]), 47

Introduction

This chapter is about the zoological "books" (chapters) by Georg Marcgraf in Piso and Marcgraf's *Historia Naturalis Brasiliae* (henceforth HNB) and more specifically his chapter on fish, with some occasional extrapolations to his chapters on other animal groups. Marcgraf's ichthyological texts and illustrations will be addressed as well as their sources, their role in the transmission of knowledge, their editing performed by Johannes de Laet (1581–1649), and their transnational reception by some famous ichthyologists: the seventeenth-century English naturalists Francis Willughby (1635–1672) and John Ray (1627–1705), the German ichthyologist Marcus Elieser Bloch (1723–1799), and finally the French zoologist Georges Cuvier (1769–1832). Their reactions make us aware of what is new in Marcgraf's ichthyological chapter.

This chapter will restrain from any further extrapolation to the botanical portion of the treatise, because this portion is quite different from the zoological one. The botanical portion counts 140 pages, whereas the zoological portion covers no more than 120 pages. Marcgraf's botanical descriptions are longer and much more detailed than the zoological ones. As for the illustrations, in his *Preface to the Reader* the editor De Laet informs us that the botanical illustrations were not made by Marcgraf but commissioned by De Laet on the basis of the specimens collected and dried by Marcgraf.[3] The botanical woodcuts are larger than Marcgraf's ones (no more than two per page, whereas the zoological pages often have three or four illustrations). Moreover, the botanical illustrations are technically more perfect and also more detailed.

DOI: 10.4324/9781003362920-7

The Two Paratexts: Practical Versus Scholarly

Our starting point will be the two paratexts of the HNB, namely Marcgraf's *Dedication* to the Brazilian governor Johan Maurits of Nassau-Siegen (1604–1679) and De Laet's *Preface to the Reader*. These paratexts give us essential information about the fabrication of Marcgraf's text. In his *Dedication*, Marcgraf presents himself as a meticulous observer, who made his own descriptions and illustrations and himself collected the names the Indigenous Brazilians gave to the animals described. This is important information, because, by doing so, Marcgraf positions himself as a "practical man" (as understood by Anthony Grafton),[4] a man of direct observation, whose perspective is not biased by any scholarly knowledge. This directness, this "autoptic imagination," has been recently addressed by Neil Safier in his article *Beyond Brazilian Nature*.[5] It can be linked to one of the best known early modern texts on Brazil: namely Michel de Montaigne's (1533–1592) chapter *Des Cannibales* in his *Essais*. In this chapter, Montaigne dwells upon the usefulness of the eyewitnesses of practical men, fulgurating against "those clever" cosmographers, who always have the tendency to embellish their reports. Montaigne eloquently exemplifies his argumentation by focusing on one particular person from his household who stayed for a long time in French Brazil: "I have long had a man with me who stayed some ten or twelve years in that other world which was discovered in our century when Villegaignon made his landfall and named it *La France Antartique* [...] That man of mine was a simple, rough fellow – qualities which make for a good witness: those clever chaps notice more things more carefully but are always adding glosses; they cannot help by changing their story a little in order to make their views triumph and be more persuasive; they never show you anything purely as it is: they bend it and disguise it to fit in with their own views [...] So you need either a very trustworthy man or else a man so simple that he has nothing in him on which to build such false discoveries or make them plausible; and he must be wedded to no cause. Such was my man; moreover on various occasions he showed me several seamen and merchants whom he knew on that voyage. So I am content what he told me, without inquiring what the cosmographers have to say about it. What we need is topographers who would make detailed accounts of the places which they had actually been to. But because they have the advantage of visiting Palestine, they want to enjoy the right of telling us tales about all the rest of the world! I wish everyone would write only about what he knows – not in this matter only but in all others."[6]

   This matter-of-fact perspective of the reliable eyewitness is visible everywhere in Marcgraf's part of the treatise: in the vocabulary, syntax, and style of his Latin, the chaotic *dispositio* (composition and structure) of the different parts of his work, and the naive crudity of most of his woodcuts. While reading the tome, one has the impression that this omnipresent reliability has been carefully cultivated – I will come back to this aspect in much more detail.

In the second paratext, the *Preface*, De Laet sketches the original state of Marcgraf's manuscripts that Johan Maurits commissioned him to publish: according to De Laet, these manuscripts were "indigested and imperfect commentaries,"[7] which for reasons not specified were written in a secret code that he had to decipher. Editing Marcgraf's text was therefore a "laborious and painstaking" enterprise.

De Laet presents himself as Marcgraf's scholarly counterbalance. This is visible in the great number of annotations by De Laet himself – annotations that give the tome the learned outlook that is expected of any serious publication in Latin.

General Structures

It is now time to turn to Marcgraf's text itself. The general structure of the zoological portion of the treatise is atypical compared to the authoritative zoological encyclopedias by Conrad Gessner (1516–1565) and Ulisse Aldrovandi (1522–1605). The order in four chapters – respectively (a) fish, (b) birds, (c) mammals and reptiles, and (d) insects – is indeed strange. This is not the order of animals in the traditional Great Chain of Being, with the insects, after the plants and the non-living material world, at the bottom of the Creation, and mankind at the top. For the first three animal groups, Marcgraf's order follows the biblical order of Genesis I: after the plants follow the creatures of the waters, the creatures of the air, and the terrestrial animals. But what about the insects as a final category? The only order which comes close to Marcgraf's is the one of the four elements (water, air, earth, and fire) as it is thematized in the contemporary zoological work by Jan Jonston (1603–1675) as reedited by Frederik Ruysch (1638–1731) in 1718.[8] Jonston's work follows the order of the elements, respectively: fish, birds, mammals, and a final category of insects and mollusks. This final category is also visible in the four emblematic albums on animals, grouped according to the four elements, by Joris Hoefnagel (1542–1601) at the end of the sixteenth century.[9] The element of fire in Hoefnagel includes, very curiously, both the insects and mankind, thus closing the Chain of Being into one album by combining in the element of fire both the lowest and the highest living creatures. And from this perspective, it is not astonishing that Marcgraf, after his part on Brazilian insects, continues with the Indigenous Brazilians.

The internal order of these four chapters is also very atypical. There is no tendency to follow the classifications given by the ichthyological and ornithological works by Pierre Belon (1517–1564), Guillaume Rondelet (1507–1566), Gessner (not his alphabetically ordered *Historia Animalium* but his other naturalist works), or Aldrovandi. These orderings were *grosso modo* tripartite: freshwater fish, saltwater fish, other *aquatilia*. By contrast, Marcgraf's ordering of birds and fish seems to be simply arbitrary. This arbitrary character could have easily been corrected by De Laet, as it was done by Piso and several other readers of Marcgraf, such as Willughby and Ray, and Ruysch. But De Laet seems to favor the arbitrary and the incongruous, probably in order to give the reader the impression of direct observation: all Marcgraf's information is given *"prout venerant ad manus,"* as it falls into his hands, in the words of De Laet in his *Preface*. The same arbitrary order, regulated (or seemingly regulated) by sole coincidence, can be found in the naturalist chapters of the two most influential books on Brazil before Marcgraf, namely André Thevet's (1516–1590) *Singularités de la France Antarctique*, and Jean de Léry's (1536–1613) *Histoire d'un Voyage Faict en la Terre du Bresil, autrement Dite Amerique*.[10] It reaffirms the autoptic character of the publication, highlighted in its paratexts.

Descriptive Rhetorics

Strangely enough, Marcgraf's descriptions of the individual animal species do not have an arbitrary structure. They implicitly follow a descriptive standard, which is an echo of the well-structured descriptions of Gessner and Aldrovandi. Marcgraf's descriptions are mostly built up in three main parts: etymology, an often meticulous morphological description, and a brief part on the animal's utility for man. Let us take a typical example: the description of the "Petimbuaba," the Bluespotted Cornetfish (*Fistularia tabacaria*) (Figure 6.1).[11]

This is how the description begins: "PETIMBUABA for the Brazilians; in Dutch Tabac-pijpe, named after its form. The fish is three to four feet long, and the body resembles that of an eel. It has a sharp-toothed mouth, whose upper jaw is shorter than the lower one." Then all measurements of the fish are given in very much detail: "the length of the beak measures six inches, and the maximum mouth opening measures up to only one inch. The head is nine inches long from the eyes down to the very tip of the beak, the width of the head behind the eyes is five inches, shrinking slowly to about three inches." The description pauses for a moment at the fish's eyes: "it has pretty big eyes, as big as a walnut, in the form of a bird feather, with a beautiful red-colored pupil, surrounded with a silver-colored dress, with small spots on the front and the back."

Then follows a very detailed description of the position, shape, size, and color of the different fins (i.e., pectoral, dorsal, ventral, and anal fins), with, of course, special attention for the fish's characteristic tail, of which the middle caudal rays are extended as a long filament: "like that of a moray, thin, round and six inches long." Then the skin is described: "the skin is as slippery as an eel's skin, liver-colored, whitish on the ventral side, and reddish at some places." Also the characteristic rows with blue, sometimes

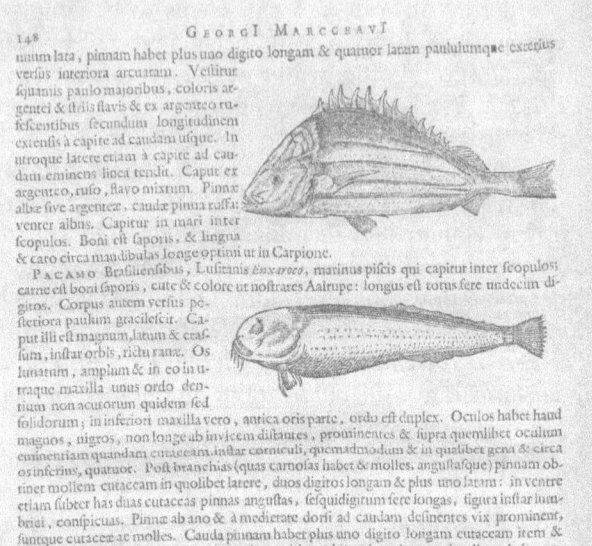

Figure 6.1 Bluespotted Cornetfish in *Historia Naturalis Brasiliae* (Piso and Marcgraf, 1648: part II, 148). Leiden University Libraries (copy THYSIA 2274).

greenish, spots over the body are mentioned. This long morphological description concludes with a short, laconic sentence: "edulis est piscis" [the fish is edible] followed by a brief reference to Conrad Gessner's work on fish: "acus piscis vocari potest de quo vide Gesnerum." [This fish can also be called *acus piscis* [thin fish]; on this, see Gessner.]

Marcgraf's description is followed by De Laet's scholarly *annotatio*: "ANNOTATIO. Aldrovandi, in his first book on fish, gives some illustrations of the *acus piscis*, but none of them corresponds to the species meticulously described by the author [Marcgraf], and whose colored drawing [of the fish] we have also seen – maybe except for Belon's *acus minor*,[12] but our species has no caudal fins such as Belon's one has."

Let us analyze this description in more detail because it is, in its rhetorical disposition, information provision, and implied ideology, typical of Marcgraf's other ichthyological descriptions. Like most of his zoological descriptions, it begins by giving the Indigenous Brazilian name. For the few descriptions where the Brazilian name is not given, this is explicitly stated (almost excused) by De Laet ("nomen Brasiliense ab Auctore non proditur").[13] In doing so, Marcgraf follows the examples of practical men, like the voyagers Thevet and Léry in their descriptions of Brazil, or Jacques Cartier (1491–1557), the French explorer of Canada, all of whom give the Indigenous names for the unknown plants and animals they encounter. This tendency can also be found in the works of the learned sixteenth-century zoologists like Rondelet, Gessner, and Belon, who, for the animals unknown to them, not only tried to forge a Latin name (or a French one, in the case of Belon) but were also interested, as true humanist scholars, in all aspects of zoological lexicology, ancient and modern, European and exotic. In the case of Marcgraf, the Indigenous name is often, but not always, followed by a Dutch name and/or a Portuguese name. In the case of the cornetfish, the Dutch name *Tabac-pijpe* appears to be a literal translation of the Tupi name, which Marcgraf probably picked up with the Dutch in Johan Maurits' court. Occasionally, in the case of the birds, a German name is given.

Marcgraf's name-giving can be interpreted diversely. First, name-giving can often be considered as an act of appropriation – and this is certainly the case here: common Dutch names, such as *baars, harde, sprot* (i.e., respectively: perch, gray mullet, sprat), imposed on exotic fish, contain an intended political message, that is, "these fish are Dutch," just as Brazil was called *Nederlands Brasilien* (Dutch Brazil). One also thinks of the provocative "Brasilia qua parte paret Belgis" as was indicated on a hand-colored map, printed in 1647, on which several Brazilian animals are to be seen, all of them taken from Marcgraf's illustrations.[14]

A second reason for giving Dutch and Portuguese names is because they are often explicative: they inform the reader on the essential physical aspects of the animals described.[15] Here are some examples of Dutch and Portuguese explicative names, to which Marcgraf added his commentary:

- "*Tabac-pijpe*, à figura dictus." [*Stem of a pipe*, so named after the fish's form.][16]
- "Belgis *een Cruyshaye*, à figura." [Named in Dutch *a cross shark* after its form.][17]

- "Piscis ingens, quem vocant *Jacob Evertsen*." [A red fish, which they [the Dutch] call *Jacob Evertsen*.][18]
- "Lusitanis *soldido* (quia armatus)." [Named in Portuguese *solid*, because it is armatured.][19]
- "*Peixe viola* Lusitanis, ob figuram quam cum cithara communem habet qua ludant Lusitani." [Named *violin fish* in Portuguese, because its form resembles a cither played by the Portuguese.][20]

This reason is also applicable to Marcgraf's bird names: these are native Dutch names, to which sometimes Dutch neologisms are added, such as "Bloemen-Specht" (Flower-Woodpecker, applied to a hummingbird), "Menscheneter" (Man-eater, applied to the urubu, a carrion bird), and "Seurvogel" (a word of unknown meaning and etymology, applied to a jabiru).[21] And, contrary to Marcgraf's fish names (which is quite understandable, because of the fish' muteness), several of his original bird names are onomatopoeias. One of them is Dutch: "Grietjebie," recorded by Marcgraf (still used today in Suriname).[22] In using (native-language) imitative bird names, Marcgraf places himself in a long tradition: beginning with the travelers Thevet and Léry, but also Belon, and later, in the eighteenth century, Georges-Louis Leclerc de Buffon (1707–1788), who was fond of native onomatopoetic name-giving – contrary, for instance, to the ornithologist Mathurin Jacques Brisson (1723–1806), who advocated a French and Latin descriptive nomenclature, i.e. name-giving that takes into account the bird's distinctive physical characteristics.[23]

There is a third reason for Dutch name-giving: it enables the intended (Dutch) reader to connect the strange and the exotic to what is familiar to them. It is noteworthy that in Marcgraf's bird chapter, Dutch is sometimes replaced by German,[24] his native language, probably for lack of adequate Dutch ornithological terminology. This explicative, "familiarizing" function of Dutch and German nomenclature is related to the only rhetorical figure of the text, namely comparison. Thus, birds are regularly compared to birds mentioned in Dutch: "Lepelaer" (spoonbill), "Meeuwe" (gull), "Kerkuyle" (barn owl), and "Waterhoen" (moorhen).[25] In the case of the "Petimbuaba," the strange fish is compared two times with the eel ("instar Anguillae"), and one time with the murine, both fish species well-known to the European reader.

This brings us to the second and main section of Marcgraf's texts, the description itself, of which we already noticed its meticulous and exhaustive character. The descriptive section is written in a dry, matter-of-fact style without any stylistic embellishment – a style that fits the practical man. As we have already noticed, the only stylistic figure allowed is the comparison, not meant as embellishment but solely serving to inform the reader about the format and the form of the fish's physical appearance: the fish is compared to an eel and a murine, and, almost poetically, its eye to a walnut and a bird feather.

Marcgraf's matter-of-fact style can also be seen in his vocabulary, and in the syntactic structure of his sentences as well. Most sentences open by mentioning the part of the body under discussion. In our example of the cornetfish: *os, caput, oculos, post branchias, post has pedes*, and *cute*. By this anteposition of the nouns, independent of their grammatical cases, the general structure of the description is made visible. In his zoological encyclopedia, Jan Jonston adopted this highlighting by syntactic anteposition and goes even further by italicizing the antepositioned elements.[26] In one of the later zoological anthologies that incorporates Marcgraf's text, namely the *Theatrum Animalium* by Frederik Ruysch, this double emphasis – anteposition and italicizing – is systematically used in the Marcgraf parts of Ruysch's book.[27]

These antepositioned nouns visualize the order of the description, which goes from head to tail, mostly ending with some general remarks on the scales or skin, especially their colors. In most of Marcgraf's ichthyological descriptions rather technical attention is given to the fins: their position (pectoral, ventral, dorsal, anal), number and form, and their constitutive rays. All this is essential for an adequate description of the anomalous fishes Marcgraf describes, such as our strangely formed cornetfish. It shows a good, practical, and up-to-date knowledge of the standards of contemporary ichthyological description, as they have been set out by Rondelet and Gessner. But it stays within the *external* morphology of the fish; no dissection is done, and there is almost no information given on the fish's *internal* morphology.

Most of Marcgraf's ichthyological descriptions (but not his description of the cornetfish) continue with a brief remark about the habitat, roughly divided into fresh water, salt water, coastal water near the rocks, and brackish water. Sometimes a more precise location is mentioned.

Marcgraf's descriptions end with a remark on the edibility of the fish described. Mostly this remark is brief – "edulis est piscis" – but regularly Marcgraf takes the opportunity to highlight the autoptic nature of his observations ("saepe commedi" [I have often eaten it]), or to give some further (literary) savoring, culinary, medical, and more spectacular details. Some examples:

- "The fish has to be roasted, because cooked he is not as good."[28]
- "The fish is not edible. But, according to the fishermen,[29] when eating it, one is paralyzed for three hours."[30]
- About the piranha: "the fish is edible; its meat is white, a bit dry, and tastes good. I have eaten this quite often." This information stands out, because in preceding it, Marcgraf brings up the danger of the piranha: "with a single bite the fish can rip off a piece of a man, as if it is cut off with a knife. Once you enter the water, even if only with a foot or hand, then you are immediately injured by this fish. The fish is fond of human blood, and loves human flesh."[31]

Generally speaking, however, the description is dry and neutral; the first person is not used, except for some rare uses of "commedi" and "vidi" [I saw,] meant to underline the autoptic mode of the description. There are some rare but interesting cases in which the author relates in the first person a personal experience with the fish under discussion, for instance the porcupine-fish: "I had a small specimen alive in February 1639, and two others in September [...] The skin like substance remains constant as it swims in the water by itself. However, as soon as the fish is taken out of the water, it turns yellow. It can blow up, and then deflate itself. It makes a sound: Uch, uch. If you want it to inflate itself, pull one of the spines on its back."[32]

Marcgraf also mentions the strange parasites he found on the Yellowtail Snapper (*Ocyurus chrysurus*). In fact, he finds these parasites so curious, that their description is almost as long as the description of the fish itself: Marcgraf even gives an illustration of these parasites. These amplifications and personal observations are rare in the fish chapter of the HNB; they are more frequent in the other zoological chapters of the treatise. This is quite understandable because birds, mammals, and insects can more easily be kept alive than fish.

Back now to our example of the cornetfish. At the end of the description, there is a brief reference to Gessner ("vide Gesnerum"). Gessner's huge folio-editions are in fact the only authority Marcgraf regularly refers to, sometimes with precise indications of chapter and pages. This makes it possible to identify the Gessner-edition he consulted, namely the very first edition of 1558.[33] In the fish chapter of his work, Marcgraf's other references, besides Gessner, are to Carolus Clusius (1526–1609) and to J.-C. Scaliger (1484–1558) – and in the bird chapter, the scarce moments when Marcgraf quotes some other sources especially concern the rather peculiar cases of the hummingbird and the bird of paradise (which does not belong to the Brazilian avifauna). I suspect these references are, at least partially, interpolations by De Laet.

De Laet's Annotations

This remarkable lack of scholarly references of course fits very well with Marcgraf's self-image as a practical man. It is in this context that the role of De Laet's annotations becomes clear. They are numerous for the fish and the mammals – less numerous for the birds. Let us take a closer look at our example of the cornetfish. As can be seen, De Laet expands Marcgraf's brief reference to Gessner, by adding the authorities of Aldrovandi and Belon, and referring to their illustrations, while explaining the differences between Marcgraf's cornetfish and the already known European species. These annotations add a learned cachet to Marcgraf's direct observations, made *sur place*, as the text wants us to believe. De Laet's references are to the whole canon of natural history: Belon, Rondelet, Aldrovandi, Clusius,

J.-C. Scaliger, with very precise indications of chapter and pages, and of course Gessner. With his references to Gessner there is something strange: all of De Laet's references to Gessner are without indication of pages. This is probably because he possessed a different Gessner edition than Marcgraf; indeed, Gessner's first edition had been sold out since 1585 and therefore it was rare in the seventeenth century – De Laet did not want to confuse the reader by referring to two different Gessner editions.[34] De Laet also quotes regularly from his own work, *Novus Orbis*, which is a commentated and illustrated anthology of the major authors on the Americas, among whom Thevet, Léry, and, very important for his annotations, Francisco Ximénez.[35] This is of course a form of self-promotion, but it also suggests the possibility of extrapolation of Marcgraf's work, as is explained in De Laet's *Preface*: Marcgraf's text should be a model to follow for new studies on other parts of America.[36] I will come back to this.

De Laet's annotations mostly serve to specify Marcgraf's observations, and sometimes to discuss and correct them. They also allow De Laet to enlarge the attractiveness of the treatise by giving further intriguing details on the animals under discussion. This is especially the case with some of Marcgraf's descriptions of birds and mammals, like the toucan, the hummingbird, and the peccary. In his annotation on the toucan, De Laet mentions a strange enormous bill of an unknown bird that he possesses in his *Wunderkammer* (which was probably of a hornbill from the East Indies). Marcgraf's description of the hummingbird leads him to discuss the presumed six-month hibernation of the bird during the rainy season. And the alleged navel on the back of the peccary makes him dissect a peccary that he received in the Low Countries.

The Illustrations

The zoological illustrations are by Marcgraf himself, and made *ad vivum*, as is explicitly said in his *Dedication* to Johan Maurits and repeated in De Laet's *Preface to the Reader*. These illustrations were colored in accordance with the very precise color-indications in Marcgraf's descriptions, as can be concluded from the brief remark made by De Laet in his annotation on the cornetfish ("we have seen Marcgraf's colored drawing of the fish").

Marcgraf's bird illustrations are in a very atypical style. In the cases where his bird illustrations are reproduced and put together with other illustrations, such as in the books by Jonston and by Willughby and Ray, Marcgraf's illustrations are immediately recognizable: his birds have a stiff attitude, staring eyes, and disproportionally small feet. Sometimes they are confusingly clumsy, as is the case with the Brazilian nightjar: the curious representation of the bird's beak is only understandable when the reader-spectator notices that the bird has its mouth wide open.[37]

By their naive crudity, the bird illustrations are the umpteenth sign of the author's autoptic perspective. However, the illustrations of the other animals (fish, mammals, reptiles, mollusks, and insects) are much less naive; in the case of the fish, they are even of high quality, as was correctly noticed by Georges Cuvier, as we shall see.

The origin of the illustrations is a complicated affaire. Marcgraf claims that he made them himself, but De Laet admits that he inserted some illustrations of his own. To be more specific, of the 86 fish illustrations, four are not made by Marcgraf: two of them are new (Mucu and Abacatuaia), the two others (Guaperua and Guaracapema) came from De Laet's *Novus Orbis*.[38] These four illustrations replace Marcgraf's own drawings, probably not for financial reasons (although reusing De Laet's two woodcuts is indeed much cheaper), but because De Laet's illustrations were simply superior. As far as I can see, in the other zoological chapters there are no other illustrations coming from De Laet's work.

A more important question is whether the watercolor drawings, conserved in the so-called *Libri Principis* (now in Kraków), often ascribed to Marcgraf, are indeed by him and were used by De Laet for the illustrations of the HNB. My provisional answer is twice "no," and by this I go against a generally accepted opinion, recently reformulated by Rebecca Parker Brienen.[39] For my argumentation it is necessary to turn to the bird illustrations and to focus on an (at first glance) insignificant detail in the description of the trogon (*Trogon spec.*).[40] Marcgraf emphasizes that the bird's feet resemble the feet of a parrot, i.e. with two toes in front and two toes behind. This observation is zoologically correct and is properly rendered in the bird's illustration (Figures 6.2 and 6.3).

However, this disposition of the toes does not correspond to the one found in the *Libri Principis*.[41] Here, the trogon's toes are clearly disposed in the "normal" but wrong way: three toes in front, one behind (see Figure 6.4).

It is noticeable that this error reoccurs at least three other times in the *Libri Principis*, namely in the case of an indefinable woodpecker, an indefinable owl, and a Spot-backed Puffbird (*Nystalus maculatus*).[42] Correctly depicted in the HNB,[43] these birds' toes are wrongly rendered in the *Libri Principis*. Moreover, in the watercolor, the trogon (which is a tree-bird and seldom sits on the ground) has been placed in such a way that its tail must hinder him.

There is a second argument that confirms my hypothesis. The trogon depicted in the *Libri Principis* has a white collar: this suggests that the species depicted is a Collared Trogon (*Trogon collaris*). But the distinctive white collar, so characteristic for this bird, is not mentioned in Marcgraf's accurate description. This implies that Marcgraf's trogon probably is another (closely related) species, namely the Blue-capped Trogon (*Trogon curucui*). This identification is supported by the trogon painted by Albert Eckhout (1610–1665) in one of his oil studies, collected in the *Theatrum Rerum Naturalium Brasiliae* (now also in Kraków),[44] which is unmistakably a blue-capped trogon,[45] depicted, by the way, with a correct position of the toes. Eckhout's watercolor bears the inscription "Curucua;" this

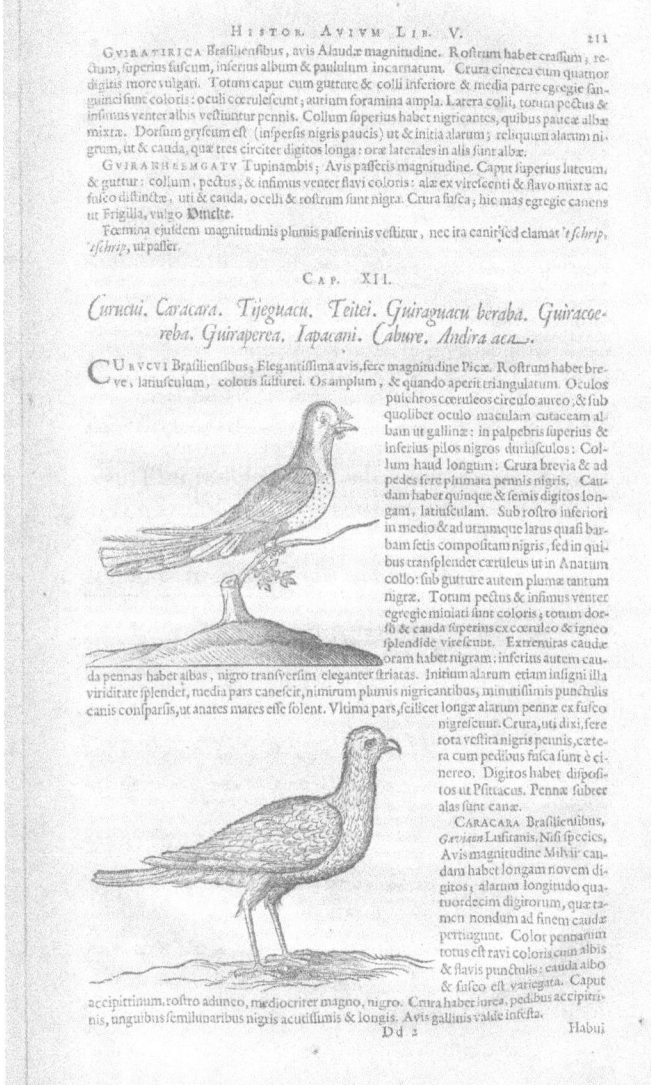

Figure 6.2 Trogon in *Historia Naturalis Brasiliae* (Piso and Marcgraf, 1648: part II, 211). Leiden University Libraries (copy THYSIA 2274).

corresponds to Marcgraf's naming of the bird: "curucui." Moreover, it appears that not only the blue-capped trogon is known to Eckhout, but also the collared trogon. Indeed, Eckhout, or someone from his direct entourage, depicted a collared trogon on one of the ceilings of the Lusthaus Hoflössnitz at Radebeul. This bird is not called "Curucua" but bears another name: "guirapotiapirangaiupar." This means that the depiction of

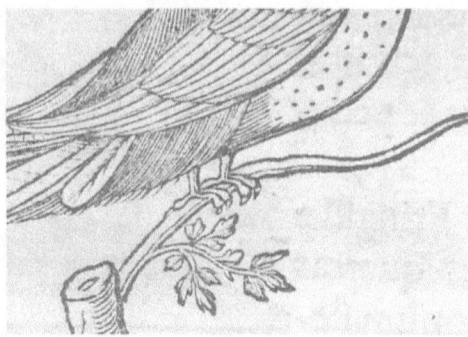

Figure 6.3 Trogon, detail of feet, in *Historia Naturalis Brasiliae* (Piso and Marcgraf, 1648: part II, 211). Leiden University Libraries (copy THYSIA 2274).

the trogon in the *Libri Principis* is doubly erroneous: not only does it have a wrong position of the toes but also a wrong name. This double mistake could not have been produced by the accurate observer Marcgraf.[46]

The erroneous identification is a fruitful one: it has been repeated in at least two hand-colored copies of the HNB, one of which is in the Leiden University Libraries (see Figure 6.5).

Figure 6.4 Trogon, watercolor, in *Libri Principis* (*Libri Picturati* A 36: f. 204). Jagiellonian Library (NDIGGRAF001151).

Figure 6.5 Trogon, colored, in *Historia Naturalis Brasiliae* (Piso and Marcgraf, 1648: part II, 211). Leiden University Libraries (copy 1407 B 3).

This implies that the hand-coloring of these copies was not based on Marcgraf's watercolor drawings (which are now lost), but on the *Libri Principis*. This corresponds with the mention of a colored copy of the tome in the auction catalogue (1668) of the colorist and "kaartafzetter" Frans Koerten (1600–1668). According to this catalogue, the coloring of this copy was modeled upon the original copy of the Prince ("volgens 't Princelijck Originael curieus afgeset") – by which undoubtedly the *Libri Principis* were meant.[47]

A third and final argument for my hypothesis is the style of the water-colors. Their coloring and uniform style do not correspond to the style of naturalist sketches, made on the spot. Nor are these watercolors meant for the engraver. Rather, they seem to be meant to be put and bound into an album on animal motifs, especially made for Johan Maurits. In this respect, this album belongs to the genre of the animal-albums, a genre that flourished since the 1560s, of which the so-called bestiary of emperor Rudolf II (1552–1612) is one of the best-known examples.[48]

A Well-Orchestrated Strategy

These illustrations make us aware of the role of the HNB in the well-orchestrated strategy of personal, political publicity, meant to confirm the merits and the achievements of the Dutch governor of Brazil, Johan Maurits, upon his return to the Netherlands. Johan Maurits could indeed easily be blamed for the rather unsuccessful Dutch colonial adventure in Brazil. In order to organize his defense, Johan Maurits made use of the most respected intellectuals. He thus appealed to the learned neo-Latin poet Casparus Barlaeus (1584–1648), who wrote a Brazilian cosmography, printed by the prestigious publishing house Blaeu.[49] Furthermore, as we noted before, a huge, hand-colored map of Brazil was printed under the title *Brasilia qua Parte Paret Belgis* (1647) by the same Blaeu. The two above-mentioned albums, the *Libri Principis* and Eckhout's *Theatrum*, were composed and Frans Post's (1612–1680) first Brazilian paintings were ordered. In their book-historical contribution to this present volume, Alex Alsemgeest and Jeroen Bos demonstrate that Marcgraf's work was pub-lished with the utmost care, without printing errors, beautifully illustrated, laid out, and printed. Addressing the tome's distribution, they demonstrate that the treatise was clearly not meant for the free market: its first pos-sessors were highly placed persons in Johan Maurits' network, as well as important scientific institutes. The more important they were, the more they were entitled to receive a hand-colored copy – this explains the pres-ence of such a copy in the Leiden University Libraries.

Legacy and Evaluations

The reception of Marcgraf's work is so widespread and long-lasting that it is not possible to do justice to it in the limits of this brief chapter.[50] I want to limit myself to some of its zoological aspects addressed above, in order to see how they have been evaluated by some of his most famous readers. Some of these readers literally copied the individual descriptions in Latin, and in the case of the *Ornithology* (1678) by Francis Willughby and John Ray, all the bird descriptions are literally translated into English.[51] Marcgraf's readers remain silent on his atypical *dispositio* and his lack of classification. Frederik Ruysch, in his *Theatrum*, follows the general structure he found in both Jonston and Marcgraf, based on the four elements, but he tries to bring in some order in the classification of the individual Brazilian species described by Marcgraf. Willem Piso, who was primarily interested in the medicinal and alimentary aspects of the animals, makes a selection of Marcgraf's animals and he also tries to order them according to a logical disposition.[52] Willughby and Ray in their books on birds and fish try to incorporate Marcgraf's fish and birds into their own sophisticated classifications.[53] From Willughby and Ray on, Marcgraf's findings were extrapolated to other parts of Latin America and beyond. In Willughby and Ray's section on the fish of the Far East, based on the findings of Johan Nieuhof, explicit references were given to Marcgraf's descriptions. In the eighteenth century, after Ruysch's *Theatrum*, Marcgraf's part of the HNB was no longer literally quoted, but incidentally incorporated in a (very) condensed form (Artedi (1705–1735) and Linnaeus (1707–1778)), or in (critical) paraphrases as did the German ichthyologist Bloch, or with stylistic embellishments as did Buffon.

More problematic were Marcgraf's zoological illustrations. Jonston, Willughby and Ray, and Ruysch were the last to produce them automatically without changes; their successors tried to find solutions for the scientific and esthetic shortcomings of the illustrations. In eighteenth-century France, in any case for the birds and mammals, they were not needed anymore, because most of Marcgraf's birds and mammals were to be seen in the enormous collection of the scientist René Antoine Ferchauld de Réaumur (1683–1757), which was later confiscated and entered into the Royal Collection. This collection found its way through the lavishly illustrated natural histories on mammals and birds by Brisson and Buffon. Marcgraf's fish illustrations remained problematic. For example, Bloch was heavily disappointed by both the technical and the ichthyological qualities of Marcgraf's woodcuts. He compared these woodcuts to the watercolors of the *Libri Principis*, which he consulted in the Royal Library at Berlin (and which, by the way, he ascribed to Johan Maurits himself). Consultation of the watercolors made him decide to turn to these albums for his description and illustration of Marcgraf's above-mentioned yellowtail snapper.[54] And he blames the bad quality of both Marcgraf's description and illustration for the fact that Artedi and Linnaeus did not incorporate the yellowtail snapper in their systems.[55] But for the cornetfish, he found not only Marcgraf's description and the woodcut insufficient,[56] but also Johan Maurits' watercolor. He therefore based himself for his illustration on three specimens of the fish he found in a collection in Leipzig.[57]

Georges Cuvier did not agree with Bloch on the scientific qualities of Marcgraf's woodcuts. In his authoritative *Tableau Historique des Progrès de l'Ichthyologie, depuis Son Origine jusqu'à Nos Jours*, he wrote: "the drawings are quite recognizable, despite the fact that they are simple wood engravings."[58] Because he felt skeptical about Bloch's interpretation of Marcgraf's illustrations, he sent his assistant and later successor Achille Valenciennes (1794–1865) to Berlin in order to copy the fishes from the *Libri Principis*, and with success. Cuvier wrote: "Valenciennes obtained permission from the conservators of the library to copy these collections, and we are today able to compare them with Bloch's copies and with nature and definitely fix the genera and species to which each fish should be referred."

With Cuvier, the zoologist turns into a comparative art historian, who is not afraid to cross national boundaries. Cuvier's interdisciplinary and transnational approach to Marcgraf can serve as a shining example for the actual historian of science – even if, in this present case, Cuvier was unaware of the unreliability of the *Libri Principis*.

Notes

1 This chapter originates from the research project *A New History of Fishes: A Long-Term Approach to Fishes in Science and Culture, 1550–1880*. This project is based at Leiden University and is funded by the Dutch Organisation for Scientific Research (NWO).

2 Georges Cuvier, *Historical Portrait of the Progress of Ichthyology, from Its Origins to Our Time*, trans. A.J. Simpson, ed. T.W. Pietsch (Baltimore, MD: Johns Hopkins University Press, 1995 [1828]), 47.

3 His herbariums are now kept in the Natural History Museum of Denmark in Copenhagen.

4 Anthony Grafton, *New Worlds, Ancient Texts: The Power of Tradition and the Shock of Discovery* (Cambridge, UK: Harvard University Press, 1992).

5 Neil Safier, "Beyond Brazilian Nature: The Editorial Itineraries of Marcgraf and Piso's *Historia Naturalis Brasiliae*," in *The Legacy of Dutch Brazil*, ed. Michiel van Groesen (New York, NY and Cambridge, UK: Cambridge University Press, 2014), 168–186.

6 Michel de Montaigne, *The Essays*, trans. and ed. M.A. Screech (London, UK: Allen Lane – The Penguin Press, 1991), 228–231.

7 Unless indicated otherwise, all translations in this chapter are my own.

8 Jan Jonston, *Historiae Naturalis [...] Libri*, 4 volumes (Frankfurt: Matthaeus Merian, 1650); ed. Hendrik Ruysch, *Theatrum Universale Omnium Animalium: Piscium, Avium, Quadrupedum, Exanguium, Aquaticorum, Insectorum, et Angium* (Amsterdam: Wetstein, 1718).

9 Recent publications on Hoefnagel include: Thea Vignau-Wilberg, *Joris and Jacob Hoefnagel: Art and Science around 1600* (Berlin: Hatje Cantz Verlag, 2017); Marissa Bass, "Mimetic Obscurity in Joris Hoefnagel's Four Elements," in *Emblems in the Natural World*, ed. Karl A.E. Enenkel and Paul J. Smith (Leiden and Boston, MA: Brill, 2017), 521–547.

10 André Thevet, *Singularités de la France Antarctique* (Paris: par Maurice de La Porte, 1557); Jean de Léry, *Histoire d'un voyage faict en la terre du Bresil, autrement dite Amérique* (La Rochelle: Chuppin, 1578).

11 Willem Piso and Georg Marcgraf, *Historia Naturalis Brasiliae: In qua non tantum Plantae et Animalia, sed et Indigenarum Morbi, Ingenia et Mores Describuntur et Iconibus supra Quingentas Illustrantur* (Leiden and Amsterdam: Elzevier, 1648), part II, 148–149.

12 Aldrovandi indeed reproduces Belon's illustration of the *acus minor*: Ulisse Aldrovandi, *De Piscibus Libri V. et De Cetis Lib. Unus*, ed. 2 (Bologna: Apud Bellagumbam, 1658), 138.

13 Piso and Marcgraf, *Historia Naturalis,* part II, 166.

14 Alexander De Bruin, *Frans Post: Animals in Brazil* (Amsterdam: Rijksmuseum, 2016), 29–35.

15 See also Willemsen in this volume.

16 Piso and Marcgraf, *Historia Naturalis*, part II, 148.

17 Ibid, 181. This fish is called "hamerkophaai" in modern Dutch, and "hammerhead shark" in English.

18 Ibid, 143; see also Ibid, 169. This is the first mention in ichthyology of the name Jacob Evertsen. Jacob Evertsen was a historical figure: a sailor, a man of little stature with a pockmarked face. See L.B. Holthuis, who found this ironic name in Jacobus Bontius, "Historiae Naturalis & Medicae Indiae Orientalis Libri Sex," in *De India utriusque Re Naturali et Medica, Libri Quatuordecim Quorum Contenta Pagina Sequens Exhibit*, Willem Piso (Amsterdam: Elzevier, 1658), 1–226. See Lipke Bijdeley Holthuis, "Who Was Jacob Evertsen? Search for the Identity of the Godfather of Some Spotted Groupers (Pisces: Serranidae: Epinephelinae)," *Zoölogische Mededelingen Leiden* 69, no. 6 (1995): 73–78.

19 Piso and Marcgraf, *Historia Naturalis*, part II, 151.

20 Ibid. For some other examples, see 151, 152, 169.

21 Ibid, 196, 207, 200.

22 Ibid, 216. The bird in question is the Great Kiskadee (*Pitangus sulphuratus*).

23 For some examples, see Paul Smith, "On Toucans and Hornbills: Readings in Early Modern Ornithology from Belon to Buffon," in *Early Modern Zoology: The Construction of Animals in Science, Literature and the Visual Arts*, ed. Karl Enenkel and Paul Smith (Leiden and Boston: Brill, 2007), 111–112.

24 "Ein Trostel" (Piso and Marcgraf, *Historia Naturalis,* part II, 208); "Gympel" (Piso and Marcgraf, *Historia Naturalis,* part II, 215).

25 Piso and Marcgraf, *Historia Naturalis,* part II, 204, 205, 205, 209.

26 Jonston, *Historiae Naturalis.*

27 Ruysch, *Theatrum Universale.*

28 Piso and Marcgraf, *Historia Naturalis,* part II, 157.

29 For the importance of Marcgraf's continuous reference to Indigenous knowledge, see Singh and Françozo in this volume.

30 Piso and Marcgraf, *Historia Naturalis,* part II, 152.

31 Ibid, 164–165.

32 Ibid, 159.

33 Conrad Gessner, *Historia Animalium Liber III: Qui Est de Piscium et Aquatilium Animanium Natura* (Zürich: Froschauer, 1558).

34 Urs B. Leu, *Conrad Gessner (1516–1565): Universalgelehrter und Naturforscher der Renaissance* (Zürich: Verlag Neue Zürcher Zeitung, 2016), 225.

35 [Dutch edition:] Johannes de Laet, *Nievve Wereldt ofte Beschrijvinghe van West-Indien* (Leiden: Elzevier, 1625); [Latin edition:] Johannes de Laet, *Novus Orbis seu Descriptionis Indiae Occidentalis* (Leiden: Elzevier, 1633); Francisco Ximénez, *Quatro Libros de la Naturaleza, y Virtudes de las Plantas, y Animales que Estan Recevidos en el Uso de la Medicina en la Nueva Espana* (Mexico: Diego López Dávalos, 1615).

36 De Laet's strategy of extrapolation and its success in eighteenth-century Europe have been convincingly addressed by Safier, "Beyond Brazilian Nature."

37 Piso and Marcgraf, *Historia Naturalis,* part II, 195.

38 Ibid, 161: *Mucu* and *Abacatuaia*; the two others are to be found in Ibid, 150: *Guaperua,* and Ibid, 160: *Guaracapema.*

39 Rebecca Parker Brienen, "From Brazil to Europe: The Zoological Drawings of Albert Eckhout and Georg Marcgraf," in Enenkel and Smith, *Early Modern Zoology,* 273–314.

40 Piso and Marcgraf, *Historia Naturalis,* part II, 211.

41   My references are to the facsimile-edition: eds. Cristina Ferrão and José Paulo Monteiro Soares, *Brasil-Holandês / Dutch-Brasil,* 5 volumes (Rio de Janeiro: Editora Index, 1995), vol. I, 63.

42   Ferrão and Soares, *Brasil-Holandês,* vol. I, 55, 89; see also Ferrão and Soares, *Brasil-Holandês* vol. II, 42.

43   To be exact: the woodpecker is well portrayed by Marcgraf. The owl's left paw is depicted wrongly, but its right paw is correct. With the tailless puffbird, which is also depicted by Frans Post, something strange is going on. Whereas, in reality, the bird has two toes forward and two toes to the back, Marcgraf portrays the bird with four toes forward (Piso and Marcgraf, *Historia Naturalis,* part II, 208). By doing so, he shows his observational qualities: he is well aware that this specimen is malformed, not only by its lacking tail ("Caret cauda"), but also by the atypical position of its toes. All other artists, including Frans Post and the anonymous artist of the *Libri Principis,* portray the bird "normally," with one toe backward and three toes forward. For some of these "normalized" depictions of the bird, see De Bruin, *Frans Post,* 39.

44   Eds. Cristina Ferrão and José Paulo Monteiro Soares, *Theatrum Rerum Naturalium Brasiliae, Brasil – Holandes; Dutch – Brazil (Icones Aquatilium / Icones Volatilium)* (Rio de Janeiro: Editora Index, 1993), 148.

45   The scientific name of the blue-capped trogon – *Trogon curucui* – is based on the Indigenous name, which imitates the call of the bird. This name seems to better imitate the call of the blue-capped trogon than the Indigenous name of the collared trogon.

46   On the bird ceiling of Hoflössnitz, see: Peter Mason, "Eighty Brazilian Birds for Johann Georg," *Folk. Journal of the Danish Ethnographic Society* 43 (2001): 103–121; Dante Martins Texeira, "Os Quadros de Aves Tropicais do Castelo de Hoflössnitz na Saxônia e Albert Eckhout (ca. 1610–1666), Artista do Brasil Holandês," *Revista do Instituto de Estudos Brasileiros* 49 (2009): 67–90, doi: 10.11606/issn.2316-901X.v0i49p67-90.

47   *Catalogus van een Menighte Treffelijcke Boecken [...] Naergelaten by Wijlen Frans Koerten [...]* (Amsterdam: Jacob Lescailje, 1668), 4, lot 40; cf. Alsemgeest and Bos in this volume.

48   Marrigje Rikken, "Dieren Verbeeld: Diervoorstellingen in Tekeningen, Prenten en Schilderijen door Kunstenaars uit de Zuidelijke Nederlanden tussen 1550 en 1630" (PhD diss., Leiden University, 2016), 111–144.

49   Caspar Barlaeus, *Rerum per Octennium in Brasilia et Alibi Nuper Gestarum sub Praefectura Illustrissimi Comitis I. Mavritii, Nassoviae, &c. Comitis,: Nunc Vesaliae Gubernatoris & Equitatus Foederatorum Belgii Ordd. sub Auriaco Ductoris, Historia* (Amsterdam: Joan Blaeu, 1647).

50   For some examples of Marcgraf's eighteenth-century readership, see Safier, "Beyond Brazilian Nature."

51   Francis Willugby and John Ray, *The Ornithology* (London, UK: Printed by A.C. for John Martyn, 1678).

52   Willem Piso, *De Indiae utriusque.*

53   Francis Willugby and John Ray, *De Historia Piscium Libri Quatuor* (Oxford, UK: E Theatro Sheldoniano, 1686).

54   "Dem Marcgrav haben wir die erste Bekanntmachung dieses Fisches zu verdanken, aber seine Zeichnung ist bey weiten nicht so gut, als die des Prinzen Moritz, die ich hier mittheile." Marcus Elieser Bloch, *Naturgeschichte der ausländischen Fische,* vol. V (Berlin: bey den Königl. Akademischen Kunsthändlern J. Morino & Comp, 1791), 30.

55   "Ohne Zweifel sind die nicht genug karakteristischen Beschreibungen und schlechten Abbildungen des Marcgrav und Piso schuld daran, dass Artedi und Liné diesen Fisch in ihre Systeme nicht aufgenommen haben." Bloch, *Naturgeschichte,* 30.

56 "Marcgraf hat die Zeichnung des Prinzen Moritz schlecht kopirt, und da er sel-
    bige bei seiner Beschreibung zu Grunde legt, so hat diese nichts anders als unge-
    treu ausfallen können. ... Von Piso an haben alle Naturkundiger der Irrthum des
    Marcgraf fortgepflanzt". Bloch, *Naturgeschichte*, 128.
57 "Handzeichnung, Bloch, Fiscularia tabacaria ... tabacaria," *Humboldt-Univer-
    sität zu Berlin: Lautarchiv*, accessed 30 May 2022, https://www.lautarchiv.
    hu-berlin.de/objekte/historische-arbeitsstelle/20881/.
58 Cuvier, *Historical Portrait*, 47. Mariana Françozo draws my attention to Cuvi-
    er's personal copy of Marcgraf's *Historia Naturalis Brasiliae*, with Cuvier's
    annotations; previously listed for sale on the website of Arader Galleries; listing
    taken offline before May 2022.

7 Reconnecting Knowledges

Historia Naturalis Brasiliae back to Indigenous Societies[1]

Aline da Cruz and Walkíria Neiva Praça

Introduction

In the popular imagination, the "indigenous" is seen as a homogeneous group of people, without any regard to their cultural differences – let alone linguistic ones – consisting of over 254 Indigenous groups living in the Brazilian territory.[2] In addition, history taught in Brazilian schools, and in other parts of the world, completely conceals Indigenous knowledge, removing each of these groups from their mythologies and their millennial knowledge about astronomy, agriculture, medicine, etc. As observed by Pereira,[3] since the *Carta de Caminha*, "indigenous" peoples are represented as a type of white board onto which the colonizer's mark could be printed:

> Parece-me gente de tal inocência que, se nós entendêssemos a sua fala e eles a nossa, seriam logo cristãos, visto que não têm nem entendem crença alguma, segundo as aparências. E, portanto, se os degredados que aqui hão de ficar aprenderem bem a sua fala e os entenderem, não duvido que eles, segundo a santa atenção de Vossa Alteza, se farão cristãos e hão de crer na nossa santa fé, à qual preza a Nosso Senhor que os traga, porque certamente esta gente é boa e de bela simplicidade. E imprimir-se-á facilmente neles qualquer cunho que lhe quiserem dar, uma vez que Nosso Senhor lhes deu bons corpos e bons rostos, como a homens bons.
>
> [For me they seem such of naive people that, if we could understand their speech and they could understand ours, they would soon be Christians, as they don´t have nor understand any belief, according to appearances. And, therefore, if the degraded, who shall stay around, learn their speech well enough and make themselves understood, I believe they, following the attention of your highness, will become Christians and will believe in our Faith, which cherishes to Our Lord to bring them, because they are certainly good people and people of beautiful simplicity. And it will be easy to imprint upon them any stamp they would give, since Our Lord has given them good bodies and good faces, as for all good men.][4]

DOI: 10.4324/9781003362920-8

The Portuguese representation of Indigenous people creates an idyllic imaginary in which man establishes a completely peaceful relation with nature. In this representation there is no space for Indigenous conflicts, nor for their way of acting in the world and their knowledge. As has been well observed by Lemos Barbosa, there was no interest in registering Indigenous narratives:

> Os antigos missionários pagaram tributo à mentalidade dominante da época. Considerando a cultura europeia e as línguas clássicas o tipo ideal de cultura e de linguagem humanas, mas não lograram compreender o interesse de registrar produções espontâneas de uma língua de índios. Deixaram-nos inúmeras traduções de livros europeus, de composições ocidentais; não nos legaram uma só lenda ou narração autêntica no idioma nativo.
>
> [The ancient missionaries paid tribute to the prevailing mindset of the time. Considering European culture and classical languages the ideal type of human culture and language, but failed to understand the interest of recording spontaneous productions of an Indian language. They have left us numerous translations of European books, of Western compositions; they have not bequeathed us a single legend or authentic narrative in the native language.][5]

In contrast to the poetic view of the Portuguese documentations, the documentation produced during the period of Dutch Brazil (1630–1654) represents Indigenous people with their knowledge of Brazilian nature. With financial support from Johan Maurits of Nassau-Siegen a series of documents were produced that described Brazilian nature *in loco*: the codices *Theatrum Rerum Naturalium Brasiliae*, *Libri Principis*, and *Miscelânia Cleyeri*. In these three codices more than 900 animals and plants are represented by drawings made by Eckhout, Marcgraf, and, at least, three other artists.[6] Together with the drawings there is also information based on three main criteria: the possibility of eating it, uses in daily life (such as transport, work, and clothes), and the possibility of domestication.[7]

Back in the Netherlands, the results of this project of documenting Brazilian natural history were used to elaborate the treatise *Historia Naturalis Brasiliae* (henceforth HNB) by Willem Piso and Georg Marcgraf, published in 1648 in Leiden and Amsterdam by Elzevier.[8] The treatise is a record of Brazilian vegetation and wildlife along with detailed descriptions, uses, and the terminology used to name them (and classify them) in the language that is conventionally called "Tupi." The documentation registered in the HNB is not restricted to natural history: there are also descriptions of the Indigenous peoples of Brazil.

According to Whitehead and Boeseman,[9] the HNB became an important source on the wildlife and vegetation of South America, evidenced by the fact that Carl Linnaeus used Tupi terminology and the descriptions of species to create his taxonomic system in the second half of the eighteenth century. Because of this, Tupi terminology gained the status of appearing in scientific terms, such as *Passiflora murucuja* or *Bothrops jararaca*. Moreover, loanwords from Tupi are found in languages that have never established direct contact with it, such as French (Fr), English (En), and Dutch (Nd). For instance, there are loanwords in these European languages for the names of plants, such as *acajou* (Fr) and *cashew* (En) from [akaju], *manioc* (Fr/En) from [mani'ʔok], and *tapioca* (Fr/En) from [tɨpɨ'ʔok]; as well as for the names of animals, such as *ara* (Fr) or *arara* (Nd) from [a'ɾaɾ], *agouti* (Fr/En) from [aku'ti], *jaguar* (Fr/En/Nd) from [ja'war], *paca* (Fr) from ['pak], *tapir* (Fr/En/Nd) from [tapi'ʔir]).[10]

In light of this, we should ask: how was the knowledge documented in the HNB originally produced? How did the naturalists Willem Piso and, mainly, Georg Marcgraf gain access to information on the names and uses of plants and animals? As highlighted by Françozo,[11] in the literature, little attention is given to the fact that the knowledge provided in the HNB was collected among Indigenous and local peoples. Only Indigenous guides could know the Tupi terms and the local uses of Brazilian plants and animals. Most of the knowledge registered in the HNB is, therefore, Indigenous knowledge. However, the Indigenous protagonism in the production of this knowledge is hidden, since they are merely seen as informants.

Nevertheless, even if the Indigenous participants in the production of the treatise are unknown, the knowledge registered in the HNB can be seen as the immaterial patrimony of the Indigenous descendant communities. However, the existence of the tome is completely unknown to the Indigenous societies whose ancestors provided the information about the species registered in the treatise.

In order to reconnect Indigenous communities with the knowledge documented in the HNB, the project *Reconnecting* was created with the basic idea of presenting the HNB to Indigenous communities in Brazil; more specifically, Apyãwa, Baré, and Tapeba groups. In this chapter, we present a discussion on the preservation of the Tupi terminology among these peoples.[12]

This chapter is organized as follows: we start by presenting a discussion about who provided the Indigenous knowledge registered in the HNB, focusing on the relation between the Tupi and Tapuia Indigenous groups and on the concept of "the Tupi language." Then, we provide information on the people chosen to participate in this project and we briefly explain the methodology. In the next section, we compare Tupi terms registered in the HNB to the terms used by the Apyãwa, Baré, and Tapeba. Finally, we set out some considerations on the data provided and suggest future work perspectives.

The Tupi Language as a Hint to Find the HNB's Hidden Indigenous Authors

In the HNB, plants and animals are presented with terminology in a language called "brasilians" in contrast with "lusitanis" terminology. As is well known in the literature, the term "brasilians" corresponds to the language conventionally called Tupi, or even Old Tupi. Meaning that, even though there were about 1,500 different languages in the Portuguese territory in South America,[13] only one language was chosen to represent Brazilian Indigenous knowledge: Tupi, from Tupi-Guarani, a branch of the Tupian family, which was used in the colonial period as *língua geral* (general language).

The term "língua geral" is used in colonial documentation to designate Indigenous languages chosen by the colonial administration and the church as an interethnic language.[14] In Spanish America, the chosen languages were the ones used in pre-colonial Empires for interethnic communication and political domination. That was the case of Nahuatl, used by the Aztec in the vast Empire of Tuxpan, in the Atlantic Ocean, to Cihuatlan, in the Pacific Ocean, and extending from the territory where Mexico City is today, to the south of Mexico, in the provinces of Oaxaca, Guerrero, Istmo de Thuantepec, Chiapas, and until Guatemala. Similarly, Quechua had a fundamental role in the Inca Empire, along the west coast of South America from Ecuador until north of Chile, passing through the high mountains of Peru, Bolivia, and Argentina.

In Portuguese America, there was no pre-Colombian empire; in other words, no Indigenous society which politically and militarily dominated other Indigenous societies. However, according to rich colonial documentation, when the Portuguese arrived on the coast of Brazil in the sixteenth century, they found a language widely used along much of the Brazilian coast, as indicated by the Jesuit José de Anchieta:

> Desde o rio Maranhão, que está além de Pernambuco para o norte, até a terra dos carijós, que se estende para o sul, desde a Lagoa dos Patos até perto do rio que chamam de Martim Afonso, em que pode haver 800 léguas de costa, em todo sertão dela que se estenderá com 200 ou 300 léguas tirando o dos carijós, que é muito maior e chega até as serras do Peru há uma só língua.
>
> [From the river Maranhão, which is beyond Pernambuco to the north, where you will find the land of carijós, which extends southwards, from Lagoa dos Patos to near the river they call Martim Afonso, in which there may be 2400 miles of coast, with all of its backwoods extending over 600 or 900 miles, not considering the carijós, which is much larger and reaches the hills of Peru, there is a single language.][15]

Not only in the religious documentation but also in the laity chronicles there are mentions of a language widely used throughout the coast of

Brazil, as the chronicler Gabriel Soares de Souza indicates in his *Tratado Descritivo do Brasil* of 1587:

> Embora os tupinambás estejam divididos em grupos, que são inimigos um do outro, falam a mesma língua que é quase geral na costa do Brasil.
> [Although the tupinambás are divided into groups, who are each other's enemies, they speak the **same language** which is **almost general** along the **coast of Brazil**.][16]

Similarly, the Dutch chroniclers mention this language of general use, identified on the coast of Brazil, as the language spoken by the Tupinambás, Tobajaras, and Potiguaras peoples. For this reason, this language came to be known in the literature as "Tupi," "Old Tupi," or "Tupinambá."

> Os nativos do Brasil estão agrupados em diferentes nações, que se distinguem por seus nomes: Tupinambás, Tobajaras, Petiguarás, Tapuias, Tapuyers ou Tapoeyers. As três primeiras nações usam a mesma língua que difere apenas em dialetos. No entanto, este último é dividido em várias tribos que estão distantes tanto em costumes quanto em línguas.
> [Brazilian natives are grouped in different nations, which are distinguished by their names: **Tupinambás, Tobajaras, Petiguarás,** and **Tapuias, Tapuyers, or Tapoeyers**. The first three nations use the same language that differs only in the dialects. However, the latter is divided into several tribes that are distant both in customs and in language.][17]

As Rodrigues observed, this language was "highly functional for those who wished to extract pau-brasil and settle along the coast: learned from one point of the coast, it allowed the communication in almost any other point."[18] Thus, given its functionality for colonial administration and for catechesis, this language of widespread use was also the most well-documented Indigenous language in the colonial period, with emphasis on the *Arte de Gramática da Lingoa mais Usada na Costa do Brasil* by Father José de Anchieta.[19] According to Rodrigues,[20] the Jesuit would have written a grammar of Tupiniquim, used in São Vicente around 1560, however, after visiting other colonial regions (Rio de Janeiro, Espírito Santo, and Bahia), the missionary would have perceived a broader use of Tupinambá as compared to Tupiniquim. From this observation, the missionary would have revised the manuscript in order to focus not on Tupiniquim but on Tupinambá. In the seventeenth century, Tupinambá was also described by Luís Figueira in his *Arte da Língua Brasílica*, published in Lisbon in 1621.[21] We should also mention the *Vocabulário na Língua Brasílica*, an anonymous work of the same year that would have been elaborated by Catholic missionaries.[22]

The fact that the language used to register Indigenous knowledge in the HNB was Tupi does not necessarily imply that only Tupi speakers' communities were involved in the production of the treatise. The Dutch and the Tupi communities were also in contact with Tapuia groups, and, thus, the document may also register knowledge of other Indigenous groups, whose languages were not privileged.

The Peoples and Their Trajectories

The existence of a register on Indigenous knowledge of plants and animals, with information on their uses and a Tupi terminology, should be seen as Brazilian Indigenous communities' immaterial patrimony. From this perspective, this preliminary research intended to present the HNB to three Indigenous peoples: the Baré, the Apyãwa, and the Tapeba. The Apyãwa live in a transition territory between the Amazon rainforest and the cerrado (an environment similar to African savannas). The Baré live in the Amazon rainforest. Lastly, the Tapeba live along the coast of Brazil, where there is a predominance of dry vegetation, called "caatinga,"[23] the same territory where the majority of data registered in the HNB was collected (Figure 7.1).

Apyãwa

The Apyãwa, traditionally known as Tapirapé, are approximately 1,100 people, living in two Indigenous areas, known as the Tapirapé/Karajá Indigenous Land and the Urubu Branco Indigenous Land, called Tãpi'itãwa, located in the northeast of Mato Grosso, in Brazil. According to Wagley,[24] they are an Amazonian people, adapted to the humid tropical rainforest, sharing their ways of life with other native peoples of the Amazonian hydrographic system. Although they are located in central Brazil, the Tapirapé, as well as other Indigenous communities that speak Tupi, only migrated to this region after 1500. In addition, Wagley compares the Tapirapé to the coastal Tupinambá and demonstrates that these peoples are intimately related in both art and culture, myth, and religion.[25]

According to ethnologist Herbert Baldus, the Tapirapé have been in central Brazil for some centuries, as can be seen in Ehrenreich:

> Na região do rio Araguaia vivem os Tapirapé que, é verdade, não foram visitados por viajante algum, mas que já no século passado tiveram relações com colonos.
>
> [In the Araguaia River region live the Tapirapé, who in fact were not visited by any traveler, but who had relations with colonists in the last century.][26]

In the same way as von Martius,[27] Ehrenreich incorporates these Indigenous people into the "Central Tupi" groups. The Apyãwa language (Tapirapé),

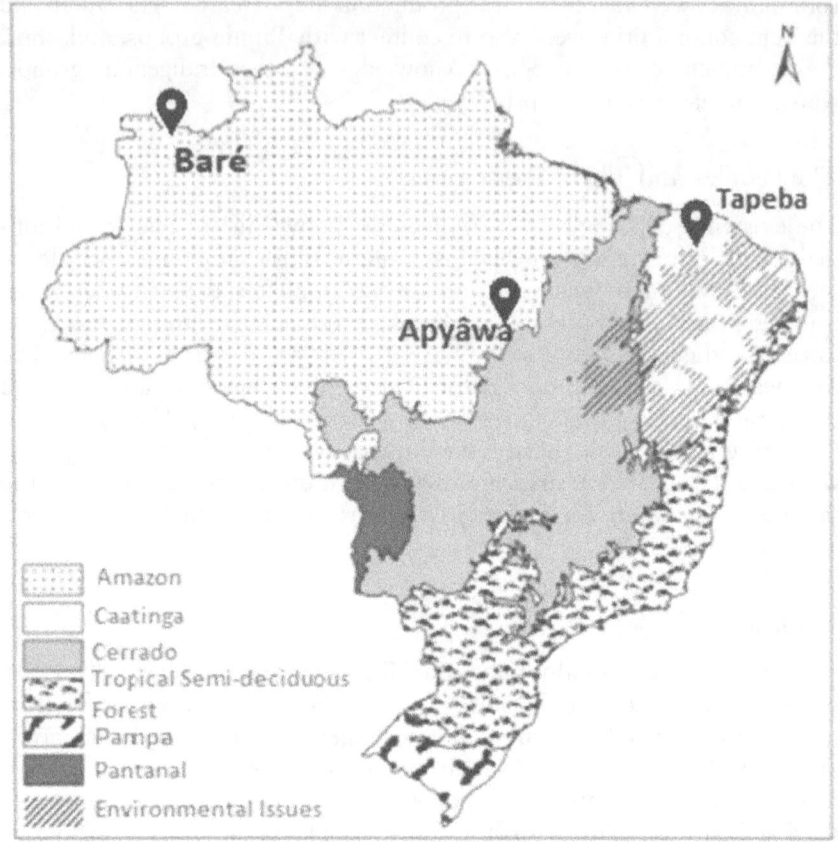

BRASIL. Environmental Ministry/IBGE. Biomas. 2004 (adapted).

Figure 7.1 Territories where the Baré, the Apyãwa, and the Tapeba live on a map of Brazil. Adapted from: Environmental Ministry/IBGE, Biomas, 2004.

used with vitality by all generations, is classified by Rodrigues and by Rodrigues and Cabral as belonging to subset IV of the Tupi-Guarani family, from the Tupi stock, which also includes Asuriní do Tocantins, Avá-Canoeiro, Guajajára, Parakanã, Suruí (Mujetire), Tembé, and Turiwára.[28]

The Apyãwa people are surrounded by Jê people, such as Yny (traditionally known as Karajá), the Au'wẽ (also known as Xavante), and Kayapó, with whom they lived in struggles over territory. However, nowadays, they live peacefully with everyone.

Even if nowadays the Apyãwa live in peace with the Jê groups, there are records of important conflicts between these groups. According to Irmãzinha de Jesus Genoveva, a Catholic missionary of the Irmãzinhas de Jesus congregation who lived and worked with the Apyãwa for over 60 years, and to Wagley,[29] due to a war against the Kayapó Metuktire in 1947, the

Apyãwa abandoned their original territory and dispersed throughout some municipalities of the Araguaia valley. In that period, the number of Apyãwa was reduced to only 47 people. At the end of 1947, Valentim Gomes, who had been a guide to Charles Wagley in 1939 and who was known to the Apyãwa, met them, with the collaboration of the Dominicans, where today the Tapirapé/Karajá Indigenous area is located. As a result of this process, the Apyãwa society was cohesively reestablished, increasing their population. It should be noted that, due to the decision of establishing the new Apyãwa territory in the Indigenous Territory Tapirapé/Karajá, the contact between Yny and Apyãwa has increased a lot and interethnic marriages between these two groups have become very frequent.

The choice of Apyãwa as participants in this project is based on the proximity between Tupinambá, the language used in HNB to register Indigenous terminology, and Apyãwa. As seen before, they are both Tupi-Guarani languages. Moreover, Apyãwa is a very conservative language in relation to other languages of the family. According to Praça, despite a significant change in its phonological system, the Tapirapé language is conservative in terms of morphology, retaining many morphemes that in other languages of the same family have already been lost.[30] Even though they speak very closely related languages and they share cultural ancestors, Apyãwa live in a very different environment compared to Tupinambá. These people used to live on the Brazilian coast, whereas the Apyãwa do not have any contact with the sea and live in the transition between the Amazonian rainforest and the Brazilian savannas.

Baré

As we saw above, the so-called Old Tupi language played an important role in the colonization of Portuguese America, thanks to its use in great territorial extension even before the arrival of the colonizers. For this reason, it was used as a language of interethnic communication. From the seventeenth century, Old Tupi began to spread within the Amazon region, undergoing profound grammatical changes, becoming a *língua geral*, which, from the nineteenth century, came to be known as Nheengatú, whose name has etymological origins in the compounding of *nheen* [language] and *katu* [good,] that is, "the good language." Today, Nheengatú is spoken in the Upper Rio Negro by Baré, Baniwa, and Werekena people who have replaced their traditional languages of the Arawak family with Nheengatú and, more recently, are undergoing a new process of linguistic substitution, since Nheengatú is being replaced by Brazilian Portuguese.

Baré was chosen to be part of this project due to the fact that it is one of these Arawakan groups who have swiched to Nheengatú, the modern variety of Old Tupi as documented in the HNB. The Baré people occupied a vast territory between Timoni Island, in the present municipality of Santa Isabel do Rio Negro, to the Casiquiare Channel, in Venezuela.[31] However, the Baré territory has decreased from the first half of the eighteenth century onward due to the policy of destocking, that is, the capture of Indigenous people from the Rio Negro to work on the extraction of wild products, and also to work on the farms in Belém and São Luís. In the nineteenth century, the Baré population replaced their own language with Nheengatú, due to the fact that the traditional Baré territory was occupied by rubber tappers who spoke Nheengatú. Thus, given the importance of Nheengatú to the economy based on the extraction of rubber, Baré gradually ceased to be spoken until its last speakers died in the late 1990s, as recorded by Aikhenvald, Oliveira, and Valteir Martins (personal communication).[32] Currently, the Baré population comprises about 8,000 people, who inhabit the Upper Rio Negro, particularly the banks of the Rio Negro and the urban area of the municipality of São Gabriel da Cachoeira. There are also Nheengatú speakers in Venezuela and in the middle and lower Rio Negro, particularly in Santa Isabel do Rio Negro and near Manaus.

Tapeba

From the earliest years of Portuguese colonization in America, northeast Brazilian coastal peoples suffered from the processes of extermination and cultural loss and transformation. In Ceará, where the Tapeba dwell, a Provincial Report of 1863 decreed that the 'Indians' in the state were extinct. In the words explicitly used in the document: "here, there are no longer indigenous villages or wild Indians."[33] In this way, the people of the Brazilian Northeast were no longer able to speak Indigenous languages as this was one of the main arguments used to deprive them of their rights to the land and made these peoples invisible to the public policies originally developed to serve them.

In addition to these linguistic aspects, the nineteenth-century report on the disappearance of 'Northeast Indians' was based on old dualistic conceptions, such as pure versus acculturated Indians, resistance versus acculturation, historical structure versus historical process. For this purist conception, the contacts established between Indigenous and non-Indigenous peoples and, particularly, the emergence of a mestizo phenotype through marriages between natives, blacks, and whites, would be proof of the extinction of the Indigenous peoples of the Brazilian Northeast. The material consequence of this purist ideology was the spoliation of the right to the land, based on the discourse that the natives would already be mixed and "civilized."

The Tapeba people, participants in this research, are an exemplary case of people invisibilized by the State and the national elites, imbued with a purist ideology, who considered themselves the holders of the power to judge who could be considered Indigenous and who could not. In the case of the Tapeba, their ethnicity was questioned precisely because they are a people formed from the conciliation between Indigenous peoples of four ethnic groups, who would have been torn by colonizing oppression, as Barreto Filho explains:

> Os Tapeba resultam de um processo histórico de inter-relacionamento e individuação étnica de segmentos de quatro povos indígenas distintos ali reunidos e vivendo sob diferentes regimes de administração de indígenas e sob diversas legislações de ordenamento fundiário ao longo do tempo: os Potiguara originários, os Tremembé, os Kariri e os Jucá – aos quais, teriam se reunidos negros libertos e/ou fugidos da escravidão.
>
> [The Tapeba are the result of a historical process of interrelationship and ethnic individuation of segments of four distinct Indigenous peoples gathered there and living under different regimes of Indigenous administration and under various land use legislation over time: the original Potiguara, the Tremembé, the Kariri, and the Jucá – with whom freed blacks and/or those escaped from slavery would have merged.][34]

Invisibilized by the Provincial Report of 1863, which decreed with force of law the non-existence of Indigenous peoples in Ceará, and by the persecution suffered from the oligarchies of the region, the Tapeba had to hide, and for that they had to abandon provisionally their more explicit cultural practices, such as the use of body painting and Indigenous clothing. In the words of Weibe Tapeba, then president of the Association of Tapeba de Caucaia Indians (ACITA), ethnic silencing was a strategy for Indigenous peoples to survive the violent repression they had suffered since the arrival of the settlers:

> No curso da história, o Povo Tapeba foi vítima de massacres e genocídios sendo o "silenciamento étnico" tapeba a arma utilizada para que o nosso povo continuasse existindo.
>
> [In the course of history, the Tapeba People were victims of massacres and genocides, and 'ethnic silencing' was the weapon used to keep our people alive.][35]

From 1980 onwards, the situation of the Tapeba began to change thanks to the pressure of popular movements now demanding the Brazilian government to recognize them as Indigenous people. It can be said that from the 1980s, the Tapeba broke with 'ethnic silencing' and began to live a period of ethnic resurgence, or ethnogenesis. At the cultural level, this process occurred through the rescue of ancient traditions, such as Toré, singing wheels, and sacred dances, which strongly mark the Indigenous peoples of the Northeast. On a material level, Tapeba communities organized themselves politically to claim ownership of their territory and to create Indigenous schools in which their knowledge is taught along with non-Indigenous knowledge.[36] Currently, the Tapeba consists of 6,000 people who inhabit the municipality of Caucaia, 20 km from Fortaleza, capital of the state of Ceará.

The choice of the Tapeba as participants in this project comes from the fact that they live in an area adjacent to the area the Dutch settled in the seventeenth century. In this way, the territory where the Tapeba live today presents the same biome represented in the HNB. Moreover, in the ethnic formation of the Tapeba there is a significant portion of Potiguaras, one of the peoples who spoke Old Tupi and a population with whom the Dutch had frequent contact.

What Is Left in the Nheengatú from the Terminology Registered in the HNB?

As a result of the study of the identification of HNB's terms with Baré, speakers of Nheengatú, we observed the preservation of a rich vocabulary as well as of animal and plants, listed in Table 7.1. The terms found in Nheengatú occur in Brazilian Portuguese, as recorded by Cunha.[37] This common lexicon is precisely the result of the linguistic contact between varieties of *língua geral* and Portuguese since the sixteenth century.

In terms of pronunciation, it is verified that the graphemes <i> and <j>, representing the phoneme / i / realized as [i] in syllabic nucleus or as [j] outside the syllabic nucleus, remained in Nheengatú. For instance, the <iararaca> in Old Tupi is performed both in Old Tupi and in Nheengatú with pronunciation [jaɾaˈɾaka], while in Portuguese the semi-vowel [j] has become a voiced post-alveolar fricative consonant [ʒ], resulting in [ʒaɾaˈɾaka]. One should also pay attention to the differences in spelling between Old Tupi and Nheengatú. In Old Tupi, which follows the orthographic pattern of Portuguese, the grapheme <c> represents the pronunciation [k] and the grapheme <ç> represents the sound [s], whereas in Nheengatú these phonemes are represented by graphemes <k> and <s>, respectively.

Table 7.1 Terms preserved in Nheengatú.

Old Tupi (as in HNB 1648)	Nheengatú		Brazilian Portuguese
çucurucu	Surukuku	[suɾukuˈku]	surucucu
boiguaçu ~ iiboya	Jibuia	[ʒiˈbuja]	jibóia
Iararaca	Jararaca	[jaɾaˈɾaka]	jararaca
Cururu	Kururu	[kuɾuˈɾu]	sapo cururu
mandihoca	Maniaka	[maniˈjaka]	mandioca
Acaju	Akaju	[akaˈju]	cajú
Inaiá	Inaja	[inaˈja]	inajá
ambaíba	Ambaiwa	[ãmbaˈiwa]	sambaíba
bacoba ~ banana	pakua ~ waria-pakua ~ pakua-puku	[paˈkuwa]	banana pacová
Tajaoba	Taiawasu	[tajawaˈsu]	taioba
Nhambi	Wãbe	[wãˈbɛ]	imbé

Table 7.2 Lexical loans from Brazilian Portuguese to Nheengatú, with respect to the vocabulary registered in the HNB (1648).

Old Tupi (as in HNB 1648)	Brazilian Portuguese	Nheengatú
Ibiboboca	lavandeira (cobra coral)	Lavandeira
Akaju	castanha	i-kastã
-----	lacre	Lakre
-----	cana brava	kana braba
Abaremo	roda-roda	roda-roda

In relation to the imbé liana, denominated in Nheengatú wãbe [wã'bɛ] and in Old Tupi <nhambi>, it is possible to raise the hypothesis that the spelling of the term in Old Tupi had been made erroneously, since the term <nhambi > is registered in other sources as "ear."

As shown in Table 7.1, many of the terms registered in the HNB were preserved in Nheengatú and the so-called 'tupinimos' were incorporated into Brazilian Portuguese as loanwords. The reverse process also occurred; that is, Nheengatú also borrowed terms from Brazilian Portuguese. Table 7.2 lists Brazilian Portuguese loans in Nheengatú to name species recorded in the first part of the HNB. A number of these terms are commonly found in other varieties of Brazilian Portuguese, such as the lavender snake (*cobra lavandeira*), the chestnut (*castanha*) and the sealing plant (*planta lacre*).

In the case of the plant named *roda-roda* (wheel-wheel) it was not possible to identify it. However, it is verified that the nomination process by Baré used reduplication and other highly iconic resources to represent the shape of the plant, which they wrap around themselves.

The process of vocabulary renewal also occurs through the use of etymological terms clearly from the Tupi-Guarani family, but not registered in the HNB. This is the case of the animal currently known in Nheengatú as *xibuipewa* [ʃibui'pɛwa], registered in the HNB as <ambuá>, and of the plant *paxiwa iwa* [pa'ʃiwa 'iwa], registered in the HNB as <iamacuru>. It was not possible to identify the names of these species in Brazilian Portuguese.

In addition to the cognate terms between Tupinambá, Nheengatú, and Portuguese, there is also a process of vocabulary renewal for loans from other Indigenous languages, as listed in Table 7.3. As Nheengatú replaced Arawak languages, it is possible that these terms may have come from the

Table 7.3 Terms in Nheengatú of non-identified etymological origin.

Old Tupi (as in HNB 1648)	Brazilian Portuguese	Nheengatú	
Ambuá	larva de fogo	aitapuru	[ajtapu'ɾu]
Icicariba	urtiga	pinupinu	[pinupi'nu]
Abaremo	roda-roda	kunipa	[ku'nipa]
Aroeira	Aroeira	kukuna	[ku'kuna]

substituted languages. However, at the current stage of this research, it is not possible to verify this hypothesis.

In addition to identifying the names of plants and animals registered in the HNB, some Nheengatú speakers also commented on the traditional practices and knowledge associated with the species in question. For example, *pinu-pinu*, a plant of the *Urticae* family, is associated both with the development of a remedy for healing the physical body and with spiritual blessing, as explained by a Baré Indigenous collaborator:

> Asui ike yamaã amu imitima, yaseruka waa kua rupi pinu-pinu. Pinu-pinu yãdarã kua. Yane tuyu ita, payé ita takua tabenzei, tauzai kua pinu-pinu tamutawariarã. Amurame sasi kua yane pira, yane akãga. Amurame maã nhaã saruwã xinga, tauzai kua pinu-pinu tabenzei, tamusasaa yane pira rese. Nhaã upiim yane pira tiarama yasaã, ma sasi yane pira rese. Ape tauzai muito kua mitima. Yariku kua rupi, sera kua pinu-pinu.
>
> Asui sawa taminhã amu tipu di pusãga. Mairame kunhã usaã nhaã sasi rame, urikuwã uiku imembira. Ape taminhã isawa xaá, tamee nhaã kunhã takitika taminhã xaá, tameẽ nhaã kunhã uurã. Asui nhaã kutara uriku usupiri taina tiara sasi kupuku ixupe.
>
> [Here we see another plant, called 'pinu-pinu.' For us, it is called pinu-pinu. Our elders, our shamans use it for blessing. Sometimes when our body is sick, or our head is hurting, or even when we got a curse, they use this pinu-pinu for blessing, they spread this plant onto our body. When they do that, we feel that the plant 'tickles' our body, it hurts our body, but then we don't feel the curse anymore. Then we use a lot this plant, this pinu-pinu.
>
> From the leaves, we make another type of medicine. When the woman feels that strong pain, just about to give birth, they prepare that tea from pinu-pinu leaves and they give it to the woman, this way the child is rapidly delivered so the woman will no longer suffer.]

The Terminology of Fauna and Flora in Apyãwa

The Apyãwa were perplexed to learn that "foreigners formerly became interested in the knowledge of the Tupi who lived by the sea," as disclosed to us by Apyãwa collaborators in personal communication. They were also impressed by the size of the HNB tome. In possession of the images of the HNB, they discussed among themselves the names of the plants and animals in Apyãwa. Possibly due to the quality of the illustrations, there were divergences among the collaborators in relation to some terms, such as, for example, a representation of a kind of yam which was recognized as *kãrã* "yam," *maxowa* or *karãxo* "a kind of yam," and also *kãã* "a yam."[38] Although the collaborators recognized that the illustration designated a yam (generic name), they were in doubt to specify the type of yam that the illustration characterized since they cultivate several varieties of this tuber.

Table 7.4 Clear cognates in Apyãwa.[40]

Old Tupi (as in HNB 1648)	Apyãwa		Brazilian Portuguese
boicininga	Majxiniga	[majʧiˈniŋa]	cascavel
çucurucu	orokoko ~ urukuku	[orokoˈko ~ urukuˈku]	surucucu
boiguaçu ~ iiboya	xowajoo ~ xowajxiga	[ʧowajˈoo ~ ʧowajˈʧiŋa]	jibóia ou jibóia branca
Cururu	kororo ~ kururu	[koroˈro ~ kuruˈru]	sapo cururu
mandihoca	mani'aka	[maniˈʔaka]	mandioca
Acaju	Ãkãxo	[ãkãˈʧo]	cajú
Inaiá	inãxa ~ myryxiryna[40]	[inãˈʧa ~ mɨrɨʧiˈrɨna]	inajá ou similar a buriti
ambaíba	ama'ywa	[amaˈʔɨwa]	ambaiba ~ ambaúba ~ embaúba
bacoba ~ banana	pako'ã ~ ka'ão	[pakoˈʔa ~ kaʔãˈo]	banana pacová ou banana brava
Acaja	Ãkãxã	[ãkãˈʧã]	cajá
Mureci	mori'i ~ mori'iyna	[moriˈʔi ~ moriʔiˈina]	murici ou similar a murici
Ananas	Ãnonã	[ãnoˈnã]	ananas
cara (ietica)	kãrã ~ maxowa ~ karãxo	[kãˈrã ~ maˈʧowa ~ karãˈʧo]	cará
caranaibam	inãxã'o ~ inaxã ~ mokãxã	[inãʧãˈʔo ~ inãˈʧã ~ mokãˈʧã]	carnauba

In a first analysis of the data, it became clear that the Apyãwa preserved most of the fauna and flora cognates registered in the HNB, as listed in Table 7.4. However, it should be clarified that the Apyãwa phonology has changed, when compared to that of the Old Tupi registered in the HNB.

Unlike most Tupi-Guarani languages, where six oral and six nasal vowel phonemes occur, Apyãwa has five oral vowels/i ɨ e a o/and five nasal vowels/ĩ ɨ̃ ẽ ã õ/. In addition, Apyãwa presents notable changes in its vowel system over that of the Proto-Tupi-Guarani (PTG). According to Lemle, the PTG vowel system had six verbal vowels/*i *ɨ *e *a *o *u/and six nasal vowels/* ĩ * ɨ * ẽ * ã * õ * ũ/. In contrast, Apyãwa presents a series of five oral vowels and their corresponding nasals. According to Soares and Leite, the changes that affected the Apyãwa phonological system are as follows[39]:

i Rising from *a to/ɨ/.
ii Rising from *ã to/ɨ̃.
iii Nasalization from *a to/ã/.
iv Lowering from *o to/a/.
v Neutralization of the contrast between *u and *o.

In this case, there is a diachronic process, in which all occurrences of the oral vowel *a from PTG have become their nasalized correspondents/ã/ in Apyãwa. The phonological change created a chain phenomenon, where

* u > o e * o > a, e * a > a. This last change led to the change * ã > i, as exemplified below:

	Proto-Tupí-Guaraní		Apyãwa	
(i)	*kutúk	>	kotók	'furar'
(ii)	*monó	>	maná	'enviar'
(iii)	*kará	>	kãrã	'fogo'
(iv)	*nupã	>	nopi	'bater'
	*ãj	>	ij	'dente'

Considering the consonants, the Apyãwa lost the *s and *d phonemes in all environments as can be verified in *orokoko ~ urukuku* "surucucu" and *mani'aka* "mandioca." The phonemes *mb and *β from Old Tupi, exemplified by the word *ambaíba*, became/m/e/w/in Apyãwa, respectively. Thus, the word *ambaíba* registered in the HNB, "ambaúba ~ embaúba ~ imbaúba," was transformed into *ama'ywa* in Apyãwa. The grapheme in the HNB representing the glide [j], as in <inaiá>, has become, in Apyãwa, the alveo-palatal/ʧ/affricated phoneme, written with <x>. Thus, we observe the difference of the cognates for the term *inaiá*, as <inaiá> [inajá] in Old Tupi, <ináxã> [inãʧã] in Apyãwa, and <inajá> [inaʒa] in Portuguese, in which a voiced post-alveolar fricative consonant [ʒ] was used.

Moreover, the current spelling of the Apyãwa language is used in this work. In this orthographic convention, the following graphemes have values other than the most usual:

i   the grapheme <x> corresponds to the alveo-palatal phoneme/ʧ/;
ii the grapheme <'> corresponds to the glottal occlusive phoneme/ʔ/;
iii the grapheme <kw> corresponds to the labio-velar/kw/occlusive phoneme;
iv the grapheme <g> corresponds to the nasal phoneme velar/ŋ/;
v   the grapheme <y> represents the central high vowel phoneme/i/.

It should be noted that the terms *ináxa, pako'ã*, and *mori'i* are clear cognates of Old Tupi and that observing the HNB illustrations could have caused some confusion in the identification of plants. Note that *inajá* is a palm tree and the composition *myyryxi-ryna* means similar to buriti, which indicates a generic identification of the species. *Ka'ão* is the designation for "banana brava," which is a plant similar to banana, or rather, *pako'ã* "banana pacova." In the same way, we can interpret the identification of *mori'i* or *mori'iyna* "murici" or "similar to murici," that is, a type of murici. The terms *maxowa* and *karax* are different types of yam, for which the collaborators do not know the respective terms in Portuguese. In addition, the collaborators identified *caranaiban* "carnaúba" as being a type of palm tree, but since this palm is not very common in the region, they identified it by exclusion as being *ina* "babaçu," sometimes *inaxã* "inaj" and sometimes *mokãxa* "macauba."

Among the terms presented in the HNB, only one loanword was verified in Apyãwa: *okomari ~ hokomari* "cobra coral." According to Praça and Ribeiro (personal communication), this is a loan from the Yny language (Karajá).[41] Due to the interethnic contact between these two groups, there are several Yny loans in Apyãwa and vice versa. However, so far, the majority are not related to fauna and flora. The few exceptions, that is, the few terms for flora and fauna from Yny loans, are species not previously known by Apyãwa, as in the case of some trees and fish from the Amazonian savannah. In the case of the designative for coral snake, as also in Nheengatú, the term for the snake is "lavandeira," a loan from Brazilian Portuguese.

The term used by Apyãwa to refer to the jararaca snake is the *majaiwa* compound, as can be seen in Table 7.5. The *majaiwa* composition is formed by the noun *maja* "snake" (generic name), juxtaposed to the stative verb *aiwa* "be bad." Very productive in Apyãwa, the composition is a process of word formation that allows for the creation of specific designations, from the junction of two or more lexical bases. In general, these formations are iconic, in which the most relevant characteristics of the elements that compose it are considered to name an entity.

In turn, the term *ayryna* is formed by a derivation, constituted of the base ay "furry animal," added with the suffix {-*ryna*} "similarity." *Ayryna*'s description is: a very hairy caterpillar. The derivation is also a very productive morphological process for amplifying the lexicon in Apyãwa. From what could be seen, the Apyãwa replaced the old terms of the HNB by other terms in Apyãwa. That is, those terms remain Tupi terms, but are not cognate to the terms registered in the HNB. In addition, these people replaced all loans from Portuguese for Apyãwa terms.

The Apyãwa have identified the illustrations for the terms *iiticucu* and *tajaoba* as being varieties of yam, while the illustration of *iacuacanga*, like kaja, is a kind of flower that originates near the Igarapés, whose name in Portuguese we do not know. As for the illustration of the term *iorupeba*, they indicated the term *yorepewa*, which could possibly be called *jurubeba* in Portuguese, but this is still a hypothesis.

Table 7.5 Other terms in Apyãwa.

Old Tupi (as in HNB 1648)	Apyãwa	Brazilian Portuguese
Iararaca	Majaiwa (cobra-ruim)	jararaca
Ambua	ayryna (animal. peludo-similaridade)	tatarana ~ taturana ou lagarta de fogo
Iiticucu	karã ~ watarõ ~ wãxara ~ karã mõ	-
Tajaoba	wãtarõ ~ kãrã ~ (wãtarõ)	taioba
Iacuacanga	Kãja	-
Iorupeba	yorepewa	jurubeba (?)

Inspired by the HNB, Koria Tapirapé, a young Apyãwa researcher, wrote an encyclopedic entry about the cassava plant (reproduced below) based on research he conducted with an elderly woman in the community. The entry records the knowledge of the people about the many culinary uses of that plant: various types of flour, *beijú* (tapioca), porridge, *kawi* (a type of porridge with a liquid consistency). Besides these culinary uses, which are well-known by the Brazilian population in general, the collaborator indicates, quite briefly, that manioc is used as a remedy to avoid heartburn. Its use occurs in a rite of passage, when the young men pass into adulthood wearing a headdress (cocar) called Akygetãra. This first entry indicates the possibility that the HNB may be used as inspiration to propose a new encyclopedia of animals and plants known by Indigenous peoples, but this time written by Indigenous people themselves in their own language.

Mani'aka

Apyãwa reka pe a'era wetepe imagyãp mani'aka, wajkyra emamywe ranõ. Apyãwa remi'oete a'era mõ mani'aka. Ymỹ raka'ẽ itori akawo emanỹt mani'aka, wajkyra magyãwa. Exanãj mĩ imagy o'iywyra ramõ, o'iatỹ ramõ ma'ea'a wawa ramõ, ipirã, temiãra, wyrã mimaka'ẽ, mimõja, mimakeka wawa ramõ. Awaxi o'i ramõ mĩ iapa ranõ. Awaxi o'i ro'õ nã emĩ awa'yao ixeakygetaxĩ ma'e remi'o xokyry ne, py'awokãja wi.

Ymỹ raka'ẽ mĩ imagy Apyãwa xokyry ramõ mani'aka ry, ixoxoka ry ikytykipyra ry. Amawot ro'õ raka'ẽ mĩ iapawo iky'yja pe. Ty pe mĩ ixaoki wowora wi ranõ.

Typy'ãka, eatykwera ne mĩ ma'ea'a wawa ramõ ireka ranõ, mexo typy'ã xaparemõ, mexo typy'ãka imamarawipyra monowi ne, eatykwera ne ranõ. Amagy mĩ kawĩ ramõ ranõ, kawĩ ramõ mĩ iapa i'awakyga, a'e rera mĩ mani'aky. Monowi kawĩ pe mĩ imagy i'awakyga ranõ. I'awakyga mĩ axaak ino'ã pe imako'iwo, a'erẽ mĩ imoãwi yropema pe. Imoapawire xowe mĩ iapa kawĩ ramõ.

Wajkyra tanã mĩ amagy to'oma ramõ, mimõja ramõ, miyra ramõ. Ma'ea'a wawa ramõ mĩ imagy a'ega ranõ, ipira re, temiara re, wyrã re.

Mani'akãwa xowe kawi ramõ xe iapapyra. Niapa xiroãwi a'ega. Akytyk mĩ iekyjta ixowi typy'ãka, imaiĩteewo ty. A'erẽ xowe mĩ imawota imana axapype.

Emanyt mĩ Apyãwa imagy mani'aka, wajkyra, mani'akãwa, wemi'o ramõ, weka pe.

[Mandioca (Manioc)

In Apyãwa culture, manioc brava and manioc cassava are used in a variety of ways. It is the typical food of the Apyãwa people. Thousands of years ago this form of the use of manioc brava has been practiced by the Apyãwa. Several types of flour are obtained from manioc, for example: puba flour, tapioca flour, and corn flour, a type of flour in which the puba flour is mixed with corn. They are consumed with meats of mammals, fish, and birds, roasted, cooked, or in moquecas. The corn meal with pepper sauce is consumed especially in a ceremony [rite of passage], in which a young boy passes into adulthood and wears a head adornment [headdress] called Akygetāra. According to our culture the corn meal with pepper is a good remedy against heartburn. In the remote epoch the Apyāwa people reused the liquid of the grated and soft cassava when squeezing the dough. It was heated and then mixed with the pepper, it would be the pepper sauce. With the liquid they also bathed, I would say as prevention against boils.

The pure tapioca beiju, the beiju mixed with peanuts, and the leftovers are also consumed with the meat. It is also important to be consumed as porridge. To be made like porridge, it is necessary to dry the manioc pulp in the sun. When making the kawĩ, which is porridge, the puba and the peanut are crushed in the pestle, soon after it is sifted and afterwards it is added with the water.

The manioc cassava, which is known as the macaxeira, is consumed as pirão, cooked and roasted. It is the mixture of meat consumption.

Sweet cassava is consumed only as porridge, it is grated to remove the fibres and liquid. When you heat the liquid it mixes with the flour.

In this way this vegetable, besides the economic value, also reflects a great Apyāwa cultural value.]

In addition to using the HNB to make an encyclopedia for adult use, the Apyāwa also suggested the use of the illustrations of animals and plants as a stimulus for children to write texts in their mother tongue, with animals as characters of fables. As an example, one can see the text written by Oparaxowi Tapirapé on catitu and cassava:

Mani'aka

Xiwā'ā a'o mani'aka. Iwyripe mi xiwā'ā i'o mani'aka. Xiwā'ā mi a'ywy-kaj mani'aka 'owejta. Xiwā'ā i'ep mani'aka re. Xiwā'ā mi axemaawā ywaywa re.

[Catitu always eats cassava. He eats the manioc underneath, making a hole in the ground. So when the catitu wants to eat cassava, he always digs the earth. He likes cassava a lot. Catitu grows eating cassava and other foods]

New Terminologies for Ancient Traditions among the Tapeba

As we saw earlier, the Tapeba people underwent a process of ethnic silencing, which forced them to erase their identities so that they would not be persecuted by the cattle ranchers. In this process, the lexicon for plants and animals was replaced by Portuguese terms, as shown in Table 7.6. Since

Table 7.6 Lexicon in Portuguese among the Tapeba.

Old Tupi (as in HNB 1648)	Variety of Brazilian Portuguese spoken by Tapeba
Boicininga	blind snake
Çucurucu	Salamander
boiguaçu ~ iiboya	boa constrictor
Iararaca	corre-canto, cipó snake
Ibiboca	coral snake or earth snake
Ambuá	taturana, fire-caterpillar or goat's hair
Moucicu	Leech
Niquî	moré, catfish
Icicariba	(not recognized)
Urucury	Baboon coconut
aroeira (termo em Português apenas)	Guarana
Araticu	Breadfruit
Pindaiba	(not recognized)
araça iba	similar to pomegranate
Bacoba	Banana
Mangaiba	mango tree
Acaja	Mom-bitch
munduy-guaçu	Jathorpha
Manaca	Chumbinho
Ananas	Pineapple

many of these animals and plants did not exist in Portugal, the terms created in Brazil tend to compare the newly designated plant or animal to better known ones among colonizers. Thus, the jararaca is compared to a vine in the term "cobra of vine;" the ambuá caterpillar is compared to "goat's hair;" araticu is seen as similar to a loaf, hence the term "bread fruit;" and the acajá happens to receive the name of mother-bitch (literally, mother-female dog). Other animals and plants are known for their better-known properties, such as ambuá – whose skin causes burning upon touch – which is called a "fire caterpillar;" the jararaca is designated by its tendency to hide in corners, hence the name "run to the sides;" and urucury is known for its slimy pulp, hence the term "baboon coconut."

The few terms of Tupi origin in the Tapeba vocabulary, reproduced in Table 7.7, are common to the other varieties of Brazilian Portuguese. As these terms were incorporated into the common lexicon of Brazilians regardless of their ethnic origin, they did not jeopardize the ethnic silencing strategy used by Tapeba to survive colonial oppression.

Although the terminology has been little preserved, the Tapeba have demonstrated to know many uses of these plants and animals since they use many of them in their medicinal practices and in the production of crafts and foods. Thus, for example, the baboon coconut is used in the production of a drink called aluá and in the production of necklaces and rings. Jenipapo is the basis of body painting used by these people as an

Table 7.7 Tupi lexicon preserved among the Tapeba.

Old Tupi (as in HNB 1648)	Brazilian Portuguese, spoken by Tapeba
Boiguaçu ~ iiboya	cobra de viado, ou jibóia
Cururu	Cururu
Mandihoca	Mandioca
Acaju	Caju
Ietaiba	Jatobá
Mureci	Murici
Caranaibam	Carnauba
Ianipaba	Jenipapo
Acajá	Saputi
Ambaiba	Torém
Guabiraba	gabiraba, guabiraba
Abaremo	Timbauba
Mureci	Murici
Cambuci	Cambuci
Iuripeba	Jurubeba

identity mark. In addition, the Tapeba recognize two types of genipap: the jenipapo bravo that can not be consumed, and the domesticated jenipapo, with which they produce a tea that serves as both an anti-inflammatory and an aphrodisiac. Some plants are used for multiple diseases; for example, saputi is used for the flu, for stomach pains, and for venereal diseases. In addition to the medicinal use of teapot plants, the Tapeba commonly use plants to produce medicinal baths. For example, the immersion of the human body into water flavored with torem is considered an important remedy against inflammation, and the gabiraba is used in the macumba for discharge baths, that is, for purification rituals.

The plant that best represents the Tapeba people is the carnauba, a palm tree that occurs in great amounts in the whole territory of this town. It is a plant of great economic importance for the Tapeba since they use it in a sustainable way for civil construction and for the production of furniture. In addition, from its straw the Tapeba make skirts, headdresses, thongs, handbags, and pichulas (a type of clothing that covers the breasts of women), and from the root medicines are made. As explained by Kennedy Itapewa, representative of the Tapeba youth association, the carnauba symbolizes the Tapeba people's enormous resistance. According to him, during long periods of drought, or even after a fire, one has the impression that the carnauba dies. However, with the rain, the carnauba regains its strength, appearing majestic again. Similarly, the Tapeba people would have disappeared during the long period of colonization, yet now resurge strong again, struggling to rescue their territory and their culture. Such a deep relationship between the Tapeba people and the carnauba is symbolically represented by the use of the traditional chaff made of straw, unlike other peoples who normally use bird-like headdresses.

Final Considerations

This chapter demonstrates the first results of a project that intends to make the treatise HNB, published by Willem Piso and George Marcgraf in 1648, accessible to Brazilian Indigenous peoples. As the various chapters highlight, the HNB is a treasure trove of information on botany, zoology, medicine, cooking, techniques, and materials available in nature for the creation of tools for fishing, hunting, and other everyday practices, and even for construction and transportation. The Indigenous knowledge presented in the treatise remains available for use by 'Western' societies, through access to the original tome in libraries scattered throughout the world, as well as thanks to two more recent editions in Portuguese.[42] Furthermore, the tome can be accessed online, such as the copy available through the Missouri Botanical Garden Library.

Although available on the internet, the HNB is in its own way not accessible to Indigenous peoples, nor is it easily accessible to most Brazilians. There are several reasons why access to the treatise is difficult, despite the apparent ease the internet provides. First, the very existence of the publication is not a well-known fact among Brazilians, much less among Indigenous peoples. Second, even reading the Portuguese version is often insurmountable for a population with very high levels of functional illiteracy. In addition, the form of the drawings, typical for seventeenth-century botany, makes it difficult for laypeople to understand them, as they are unfamiliar with this type of illustration.

For many Indigenous peoples, more so than a Portuguese version of the HNB, the treatise is of interest by inspiring the elaboration of other compendiums about the flora and fauna of Brazilian regions, but this time written by the Indigenous peoples themselves, when possible in their native languages. The entry concerning the "manioc" *maniaka*, written by Koria Tapirapé, and reproduced earlier in this chapter, is precisely such an example.

Despite the difficulties of access to the HNB, the knowledge deposited there may be an important tool for Brazilian Indigenous peoples to relearn what, through a strategy of survival based on ethnic silence, could not be taught by their ancestors. This does not mean that the knowledge preserved in the HNB is some sort of "true Indigenous knowledge," as this would institute the rhetoric of purism. On the contrary, we know that human cultures are dynamic, always in transformation, either by internal forces that cause each generation to bring innovations or by the force of contacts between peoples, intensified by the colonizing process. The existence of a 1648 record of Indigenous knowledge in an early period of colonization makes it possible to understand the dynamics of the cultural transformations which these peoples have gone through. In addition, it is up to each peoples to decide which knowledge registered in the HNB can be taken up by them and which should be kept as portraits of ancient civilizations.

Lexical transformations discussed in this chapter are exemplary of the need to understand more deeply the dynamics of linguistic and cultural contacts that occurred among the Indigenous peoples in the process of colonization. On the one hand, Tupi-speaking peoples have largely preserved the terminology for plants and animals registered in HNB, although Baré and Apyãwa live in environments quite different from that in which Dutch Brazil was established. On the other hand, the Tapeba, as part of their strategy of ethnic silencing, erased a significant part of the Tupi vocabulary, preserving only those terms that became part of the common vocabulary of Brazilian Portuguese.

Notes

1 The fieldwork for data from the Baré group was part of the project *Documentação das Variedades de Nheengatu do Amazonas*, by CNPq for the period 2013–2016. The data from Tapeba people is based on fieldwork by Aline da Cruz, as part of the project *Fortalecimento Cultural dos Povos Indígenas do Ceará, por Meio do Ensino do Nheengatú*. The data on the Apyãwa is based on fieldwork by Walkiria Neiva Praça. The study on the HNB was possible due to a scholarship offered by the Coimbra Group and Leiden University. We are grateful to Ivan da Cruz for the English version and to Mariana Françozo for her numerous and valuable comments. Unless otherwise indicated, all translations in this chapter are our own.
2 "Quem são?" *Instituto Socioambiental*, accessed 31 May 2022, https://pib.socioambiental.org/pt/Quem_s%C3%A3o.
3 Maria Luiza Scher Pereira, "O Tema do Índio e a Consciência da Nossa Diferença," *Suplemento Cultural' d'A tarde* 8, no. III (1997): 2–5.
4 Pero Vaz de Caminha, "Carta a El Rei D. Manuel (ortografia atualizada)" [orig. 1 May 1500], *Wikisource*, accessed 31 May 2022, https://pt.wikisource.org/wiki/Carta_a_El_Rei_D._Manuel_(ortografia_atualizada).
5 Antônio Lemos Barbosa, *Curso de Tupi Antigo* (Rio de Janeiro: Livraria São José, 1956), 18.
6 Rebecca Parker Brienen, "From Brazil to Europe: The Zoological drawings of Albert Eckhout and George Marcgraf," in *Early Modern Zoology: The Construction of Animals in Science, Literature and the Visual Arts*, ed. Karl Enenkel and Paul Smith (Leiden and Boston, MA: Brill, 2007), 273–316.
7 Mariana Françozo, *De Olinda a Holanda: O Gabinete de Curiosidades de Nassau* (Campinas: Editora Unicamp, 2014), 135.
8 Willem Piso and Georg Marcgraf, *Historia Naturalis Brasiliae: In qua non tantum Plantae et Animalia, sed et Indigenarum Morbi, Ingenia et Mores Describuntur et Iconibus supra Quingentas Illustrantur* (Leiden and Amsterdam: Elzevier, 1648).
9 Peter J.P. Whitehead and Marinus Boeseman, *A Portrait of Dutch 17th Century Brazil: Animals, Plants and People by the Artists of Johan Maurits of Nassau* (Amsterdam: North-Holland Publishing Company, 1989).
10 In Brazilian Portuguese, there are a number of loanwords from Old Tupi. These words are so common to Brazilian speakers that they normally do not realize that they do not have a Portuguese origin. See Wolf Dietrich and Volker Noll, "O Papel do Tupi na Formação do Português Brasileiro," in *O Português e o Tupi no Brasil*, eds. Volker Noll and Wolf Dietrich (São Paulo: Contexto, 2010), 81–103.
11 See Françozo in this volume.

12   The first results of this research were published in Portuguese as: Aline da Cruz and Walkiria Neiva Praça, "Preservação do Léxico do 'Historia Natuaralis Brasiliae' Entre os Baré e os Apyãwa," *LETRARIA* (2018): 39–54. In this chapter, we have included more data on Apyãwa and Nheengatú, as well as our research with Tapeba people.

13   Aryon Dall'Igna Rodrigues, "Línguas Indígenas: 500 Anos de Descobertas e Perdas," *DELTA* 9, no. 1 (1993): 81–103.

14   Cristina Altman, "As Línguas Gerais Sul-Americanas e a Empresa Missionária: Linguagem e Representação nos Séculos XVI e XVII," in *Línguas Gerais: Política Lingüística e Catequese na América do Sul no Período Colonial*, eds. José Ribamar Bessa Freire and Maria Carlota Rosa (Rio de Janeiro: EdUERJ, 2003), 57–83.

15   José de Anchieta in Altman, "As Línguas Gerais," 60.

16   Gabriel Soares de Souza, *Tratado Descriptivo do Brasil em 1587* (São Paulo: Nacional/EdUSP, 1938 [1587]).

17   Johann Nieuhof in Erani Stutz, "O Projeto Holandês no Brasil Seiscentista: Para uma Historiografia da Lingüística Brasileira" (PhD diss., USP, 2003), 175.

18   Rodrigues, "Línguas Indígenas," 86.

19   José de Anchieta, *Arte de Gramática da Língua mais Usada na Costa do Brasil* (São Paulo: Loyolla, 1990 [1595]), facsimile from the first edition.

20   Aryon Dall'Igna Rodrigues, "Descripcion del Tupinamba en el Periodo Colonial: El Arte de Jose de Anchieta," in *La Descripcion de las Lenguas Amerindias en la Epoca Colonial*, ed. Klaus Zimmermann (Frankfurt: Vervuert, 1997), 371–400.

21   Luís Figueira, *Arte da Língua Brasílica* (Lisboa: Manuel da Silva, [1621?]).

22   "Vocabulário na Língua Brasílica," *Boletim da Faculdade de Filosofia, Letras e Ciências Humanas* 137–138, nos. I-I (1938). This transcription was revised and confronted with the original manuscript, Ms. Fg. 3144, dated 1621, of the National Library in Lisbon by Carlos Drummond. For data collection and a critical analysis of the sixteenth-century Tubinambá related registered works, see Luciana Gimenes, "Fontes para a Historiografia Linguística do Brasil Quinhentista: Materiais de Análise," in Freire and Rosa, *Línguas Gerais*, 25–41.

23   The term *caatinga* comes from Old Tupi [ka'ʔa tĩga] "white forest".

24   Charles Wagley, *Lágrimas de Boas-vindas: Os Índios Tapirapé do Brasil Central* (Belo Horizonte: Editora Itatiaia, 1988), 49.

25   Wagley, *Lágrimas*, 50.

26   See Ehrenreich as cited in Herbert Baldus, *Tapirapé: Tribo Tupí no Brasil Central* (São Paulo: Companhia Editora Nacional / Editora da USP, 1970), 43.

27   Carl Friedrich Philipp von Martius, *Wortersammlung Brasilianischer Sprachen: Gloassaria Linguarum Brasiliensium: Glossarios de Diversas Lingoas e Dialectos, que Fallao os Indios no Imperio do Brazil* (Leipzig: F. Fleischer, 1867).

28   Aryon Dall'Igna Rodrigues, "Relações Internas na Família Linguística Tupi-Guarani," *Revista de Antropologia* 27/28 (1984/1985): 33–53; Aryon Dall'Igna Rodrigues and Ana Suelly Arruda Câmara Cabral, "Revendo a Classificação Interna da Família Tupi-Guarani," in *Línguas Indígenas Brasileiras: Fonologia, Gramática e História - Atas do I Encontro Internacional do Grupo de Trabalho sobre Línguas Indígenas da ANPOLL*, vol. 1 (Belém: Editora Universitária Pará, 2002), 327–337.

29   Irmãzinha de Jesus Genoveva, personal communication; Wagley, *Lágrimas*.

30   Walkíria Praça, *Morfossintaxe da Língua Tapirapé* (PhD diss., Universidade de Brasilia, 2007).

31   Henri Ramirez, *Línguas Arawak da Amazônia Setentrional: Comparação e Descrição* (Manaus: Editora da Universidade do Amazonas, 2001), 475.

32   Alexandra Y. Aikhenvald, *Bare* (Munich and Newcastle: Lincom Europa, 1995); Christiane Cunha de Oliveira, "Dupla Negação em Baré: Uma Explicação Diacrônica," *Revista do Museu Antropológico* 3–4, no. 1 (1999): 105–120; Valteir Martins, personal communication.

33 "Relatório do Presidente José Bento da C. F. Júnior, à Assembléia Legislativa Provincial," *Relatórios dos Presidentes da Província do Ceará* (1863). Biblioteca Pública Governador Menezes Pimentel. Núcleo de Microfilmagens.
34 Henyo Barreto Filho, *Relatório Circunstanciado de Identificação e Delimitação da TI Tapeba* (Brasília, 2006), 23.
35 Weibe Tapeba, "Terra Demarcada, Vida Garantida!," *O Povo*, 12 March 2016, accessed 31 May 2022, https://www20.opovo.com.br/app/opovo/opiniao/2016/03/12/noticiasjornalopiniao,3587433/terra-demarcada-vida-garantida.shtml.
36 Adelle Azevedo Ferreira, Artur Alves de Vasconcelos, and Marciano de Góis Moreira, *Plano de Gestão Territorial e Ambiental Indígena TAPEBA* (Fortaleza: Expressão Gráfica e Editora, 2017).
37 Antônio Geraldo da Cunha, *Dicionário Histórico das Palavras Portuguesas de Origem Tupi* (Brasília: Universidade de Brasília, 1998).
38 Yam is the common name for some plant species in the genus *Dioscorea* (family *Dioscoreaceae*) that form edible tubers.
39 Marilia Facó Soares and Yonne Leite, "Vowel Shift in the Tupi-Guarani Language Family: A Typological Approach," in *Language Change in South American Indian Languages*, ed. Mary Ritchie Key (Philadelphia, PA: University of Pennsylvania Press, 1991), 36–53.
40 Semantically, the suffix {-ryna} denotes an idea of similarity, in other words it indicates that one entity has a quality or character of being similar to the other.
41 Eduardo Ribeiro and Walkíria Neiva Praça, "Empréstimos Karajá em Tapirapé" (in preparation).
42 See Alsemgeest and Bos in this volume; George Marcgraf, *História Natural do Brasil*, trans. José Procópio de Magalhães (São Paulo: Imprensa Ofical do Estado, 1942 [1648]); Guilherme Piso, *História Natural do Brasil Ilustrada - Edição Comemorativa do Primeiro Cinqüentenário do Museu Paulista* (São Paulo: Imprensa Oficial do Estado, 1948 [1648]).

Appendix
Census of the Copies of Willem Piso and Georg Marcgraf's *Historia Naturalis Brasiliae* (Leiden and Amsterdam: Elzevier, 1648)

Alex Alsemgeest and Jeroen Bos

Australia

Sydney, State Library of New South Wales

Copy 1: Collection: David Scott Mitchell Collection. Shelf mark: DSM/ F508.8/P. Binding: Eighteenth-century parchment, spine with red title label. Coloration: Black-and-white. Complete: Yes. Provenance: Bookplate: George Bennet (1804–1893 | Australian physician and naturalist, fellow of the Royal College of Surgeons, England), pasted over old bookplate with name "Gosling." Donated to the State Library of New South Wales by David Scott Mitchell (1836–1907 | Australian book collector).

Copy 2: Collection: Richardson Collection. Shelf mark: RICHARDSON/ F287. Binding: Nineteenth- or early twentieth-century red leather, tooling on outside of binding and gold embossed decoration on inside of binding. Coloration: Black-and-white. Complete: No, lacks title page, illustrations have been cut from pages 141–2, 191–2, 193–4, 215–16, 223–4, 227–8, 269–70. Provenance: Donated to the State Library of New South Wales by Nelson Moore Richardson (1855–1925) and Mrs. Richardson of Weymouth, England in 1926.

Austria

Vienna, Österreichische Nationalbibliothek

Copy 1: Collection: Sammlung von Handschriften und alten Drucken. Shelf mark: BE.4.H.2.Alt-Prunk. Binding: Eighteenth-century ocher moroccan leather, supralibros of Prince Eugene of Savoy on front and back boards, marbled endpapers, edged gilt. Coloration: Black-and-white. Complete: Yes. Provenance: From the Bibliotheca Eugeniana, the private library of Prince Eugene of Savoy (1663–1736 | Austrian aristocrat, statesman, and military commander). Acquired by the library in 1736.

Copy 2: Collection: Bildarchiv und Grafiksammlung. Shelf mark: 273.802-D.FID. Binding: Half-leather with marbled boards. Coloration: Black-and-white. Complete: Yes. Provenance: From the collection of Francis II (1768–1835 | Holy Roman Emperor, from 1804 Francis I, the first Emperor of Austria). Transferred to a Fidei commiss Library in 1835. In 1921, the Fidei commiss Library was taken over as the property of the Republic and joined to the Austrian National Library. Note: Mentioned by the librarian of the emperor Peter Thomas Young in his "Schätzkatalog" from 1806 with the current number 5002.

Vienna, Universitätsbibliothek

Shelf mark: III-258.287. Binding: Contemporary parchment. Coloration: Black-and-white. Complete: Yes. Provenance: Manuscript annotation on title page: Leopoldini Societatis Jesu in Austria 1668. Library stamp: Biblioth. Universit. Vindebonensi [=Vienna University Library]. Transferred to the university library after the dissolution of the Jesuit order in 1773.

Belgium

Antwerp, Museum Plantin-Moretus

Shelf mark: Prentenkabinet B 227. Binding: Contemporary parchment. Coloration: Black-and-white. Complete: Yes. Provenance: Library stamps on title page and last page: Museum Plantin Moretus 1876. In the library of the Moretus family at least since 1675.

Brussels, Koninklijke Bibliotheek van België/
Bibliothèque Royale de Belgique

Copy 1: Shelf mark: VB 4.040 C RP. Binding: Contemporary brown calf, blind stamped with rolls and filets. Coloration: Black-and-white. Complete: Yes. Provenance: Manuscript annotation: Coll: Soc: Jesu Gand: 1648 [Collegium Societatis Jesu Ghent]. Ville de Bruxelles: library created by the French in 1795, transferred to the Ecole Centrale du Département de la Dyle (Central School of the Dyle Department) in 1797, donated to the City of Brussels in 1803. That library was acquired by the Belgian State for the Royal Library of Belgium in 1843.

Copy 2: Shelf mark: Müller 3.461 C. Binding: Contemporary parchment. Coloration: Black-and-white. Complete: Yes. Provenance: Library of Johannes Peter Müller, Berlin (1801–1858 | German anatomist). Acquired as a whole by the Royal Library of Belgium in 1861.

Ghent, Universiteitsbibliotheek

Copy 1: Shelf mark: BIB.HN.000105. Binding: Contemporary parchment, red edges. Coloration: Black-and-white. Complete: no, lacks title page. Provenance: Blue Minerva stamp. Confiscated from an unknown East Flemish monastery library by the French Revolutionary Army between 1793 and 1797. Given to the Departementale School, and consequently part of the University Library since 1818.

Copy 2: Shelf mark: BIB.HN.000251. Binding: Eighteenth-century calf, gilded edges, marbled endpapers. Coloration: Colored. Complete: Yes. Provenance: Red Minerva stamp. Manuscript annotation: ƒ56.0.0, Rooman [Gilles-Jean Rooman] (1696–1789 | Flemish magistrate and book collector). Acquired by the library in 1818 from Pierre Philippe Constant Lammens (1762–1836 | Belgian book collector). Note: This copy is mentioned in the auction catalogue: *Catalogue des livres de la bibliothèque de feu monsieur G. J. Rooman* (Gant, P.F. de Goesin, 1791), no. 1414.

Liège, Bibliothèque Universitaire

Shelf mark: R551D. Coloration: Black-and-white. Complete: Yes. Provenance: Manuscript annotation: Bibliothecae Ninivensis, dono D. Amandi Fabij Monrij SS. Cornelij et Cypriani Canonici Presbyteri [Abbey of St. Cornelius and St. Cyprianus, Ninove], 1648 or 1649. Ex libris: C. Van Hulthem [Charles (Karel) Joseph Emmanuel Van Hulthem] (1764–1832 | Belgian bibliophile).

Meise, Botanic Garden Meise

Shelf mark: VALD 90. Binding: Later red leather, decorated in gilt with rolls and filets, marbled endpapers. Coloration: Black-and-white. Complete: Yes. Note: Property of the Belgian State, on permanent loan to the Botanic Garden Meise.

Brazil

Belo Horizonte, Biblioteca da Universidade Federal de Minas Gerais, UFMG

Shelf mark: 1648 502.2 P678g. Coloration: Black-and-white. Complete: No, lacks title page. Binding: Rebound in the 1980s by the Library of the Museum of Natural History and Botanical Garden of the UFMG. Provenance: Bought by the UFMG on 20 June 1979 for the Library of the Museum of Natural History and Botanical Garden. In 2001 it was transferred to the Rare Books collection.

Brasília, Biblioteca Central da Universidade de Brasília, UnB

Shelf mark: 502(81) P678h. Binding: Contemporary parchment. Coloration: Black-and-white. Complete: Yes. Provenance: Bought at the São José Bookstore in 1963 for 120 thousand cruzeiros, probably by Ricardo Xavier da Silva, whose other books relating to Dutch Brazil were acquired at the same place and later given to the UnB library.

Campinas, Universidade Estadual de Campinas, Unicamp

Copy 1: Collection: Coleção Paulo Duarte. Shelf mark: 500.981 P676h 1648 OR/PD. Binding: Early twentieth-century leather. Coloration: Black-and-white. Complete: Yes. Provenance: Bookplate: Reformert skole i Fredericia [Reformed school in Fredericia, Denmark]. From the collection of Paulo Duarte (1899–1984 | Brazilian archaeologist and humanist). Acquired by Unicamp University at the foundation of the university in 1970, specifically bought by its first rector magnificus, Prof. Zeferino Vaz (1908–1981). Note: Annotated in at least three different hands throughout. Contains five added engraved plates.

Copy 2: Collection: Coleção Oswaldo Peckolt. Shelf mark: 500.981 P676h 1648 OR/OP. Binding: Twentieth-century black half-leather with marbled boards. Coloration: Black-and-white. Complete: Yes. Provenance: From the collection of Theodor Peckolt (1822–1912 | German naturalist, botanist, phytochemist, and pharmacist who worked in Brazil from 1847 to 1912).

Rio de Janeiro, Biblioteca Nacional do Brasil

Copy 1: Shelf mark: Livros Raros – 025A,004,002. Ex. 1. Binding: Parchment. Coloration: Black-and-white. Complete: Yes. Provenance: Library stamp Da Real Bibliotheca [Royal Library].

Copy 2: Shelf mark: Livros Raros – 025A,004,002. Ex. 2. Coloration: Black-and-white. Complete: Yes. Provenance: Ex libris: Francisco José da Serra.

Copy 3: Collection: Colecção Benedicto Ottoni. Shelf mark: Livros Raros – 025A,004,002. Ex. 3. Coloration: Black-and-white. Complete: Yes. Provenance: Ex libris: Arnold Wittens. Donated to the library by Julio Benedito Ottoni (1857–1926 | Brazilian industrialist) in 1912.

Copy 4: Shelf mark: Livros Raros – 025A,004,002. Ex. 4. Coloration: Black-and-white. Complete: Yes.

Copy 5: Shelf mark: Livros Raros – 025A,004,002. Ex. 5. Coloration: Black-and-white. Complete: Yes. Provenance: Bibl. Nac. e Publ. da Corte. [Probably Real Biblioteca Pública da Corte, the former name of the current National Library of Portugal].

Rio de Janeiro, Biblioteca do Museu Nacional, UFRJ

Copy 1: Shelf mark: In folio 168 OR. Coloration: Black-and-white. Complete: Yes.

Copy 2: Shelf mark: In folio 168 OR ex.2. Coloration: Black-and-white. Complete: Yes. Provenance: Donated by the descendants of Prof. Dr. Johann Becker (1932–2004 | entomologist of the National Museum in Rio de Janeiro) on 14 September 2007.

Copy 3: Shelf mark: In folio 168 OR ex.3. Coloration: Black-and-white. Complete: Yes. Provenance: Donated by Prof. Dr. Amélia Lúcia on 14 November 2008.

Rio de Janeiro, Fundação Fiocruz – Biblioteca de Manguinhos

Shelf mark: BR15.1 0007 1648 OBRA RARA – ARM. Binding: Parchment. Coloration: Black-and-white. Complete: Yes. Provenance: Manuscript annotation on title page: Bibliotheca Hafflighemensis [Library of the Abbey of Affligem] 1783.

Salvador, Universidade Federal da Bahia

Copy 1: Shelf mark: 58/59 H673 (LM). Binding: Parchment. Coloration: Black-and-white. Complete: Yes. Provenance: Ex libris: Le Comte de Carburil. Donated to the UFBA library by professor Frederico Edelweiss (1892–1976 | Brazilian bibliophile and specialist in Tupi-Guarani languages) in 1974.

Copy 2: Collection: Bibliotheca Gonçalo Moniz; School of Medicine. Shelf mark: OR 94(81) P678 (BGM). Binding: Rebound, modern binding. Coloration: Black-and-white. Complete: no, lacks pages 3–6. Provenance: From the collection of Dr. Egas Moniz Barreto de Aragão (1870–1924 | Brazilian medical doctor, professor at the School of Medicine of Bahia in 1911).

São Paulo, Biblioteca Mario de Andrade

Copy 1: Collection: Felix Pacheco. Shelf mark: INg 1648. Binding: Modern parchment. Coloration: Black-and-white. Complete: Yes. Provenance: Acquired by José Felix Alvez Pacheco (1879–1935 | Brazilian journalist, translator, poet, and politician) at the Maggs House in London in the 1920s.

Copy 2: Collection: Felix Pacheco. Shelf mark: LR 1 e 4. Binding: Parchment. Coloration: Black-and-white. Complete: Yes. Provenance: From the collection of José Felix Alvez Pacheco.

São Paulo, Instituto Itaú Cultural

Shelf mark: 23000822.031. Coloration: Colored. Complete: Yes. Provenance: Baron Horace de Landau (1824–1903 | French banker and collector; representative of the French house of the Rothschild family in Turin and Florence); then Collection of Joaquim de Souza-Leão (1897–1976 | Brazilian diplomat, Ambassador in the Netherlands in the 1950s); subsequently acquired by Banco Itaú in 2002.

São Paulo, Universidade de São Paulo

Copy 1: Collection: Biblioteca Brasiliana Guita e José Mindlin. Shelf mark: M1v 1258. Binding: Mottled calf, spine with six raised bands, ligature Æ in six compartments. Gilded supralibros on front cover with motto "non est mortale quod opto." Coloration: Black-and-white. Complete: Yes. Provenance: Ex libris: Rubens Borba Alves de Moraes (1899–1986 | Brazilian librarian, historian, and bibliophile). Donated to the University of São Paulo by José Mindlin (1914–2010 | Brazilian journalist and bibliophile) in 2006.

Copy 2: Collection: Museu Paulista. Shelf mark: OR 0598. Coloration: Black-and-white. Complete: Yes. Note: Label with handwritten text in ink reads: "Historiae Naturalis e Medicae Indiae Occidentalis, Amsterdam, 1658", the 5 has been crossed out and substituted with a 4.

Canada

Montréal, McGill University Library

Location: Osler Library. Shelf mark: Folio P678h 1648. Coloration: Title page colored; other illustrations black-and-white. Complete: Yes. Provenance: Manuscript annotation: Jabez Cay (1666–1703 | medical doctor in Newcastle upon Tyne), 16 shill., Lugd. Bat, 3-12-1687.

Montréal, Université du Québec

Shelf mark: LAR F 1648 QH117. Binding: Rebound in the 1980s. Coloration: Black-and-white. Complete: Yes. Provenance: Stamp on title page: Bibliothecae Majoris Collegii S.J. Sae Marie Marianapoli. Stamp on title page: Ex Bibliotheca J. Richard (first half nineteenth century | French medical doctor).

Québec, Bibliothèque de l'Université Laval

Shelf mark: QH 41 P678 1648.

Czech Republic

Prague, Národní knihovna České republiky

Copy 1: Shelf mark: 16 A 000012. Binding: Contemporary blind-stamped parchment with panel design over wooden boards. Coloration: Black-and-white. Complete: Yes.

Copy 2: Shelf mark: 18 A 000142. Binding: Brown leather, endpapers decorated with geometrical patterns in green, red, and gold. Coloration: Black-and-white. Complete: Yes. Provenance: Manuscript annotation on folium *2: [Illegible name].

Denmark

Copenhagen, Royal Danish Library

Copy 1: Shelf mark: 15, 458 02542. Binding: Green morocco with gilt decoration, stamps à petits fers, presumably bound in The Hague. Coloration: Colored. Complete: Yes. Provenance: In possession of the Royal Library since 1654. Donated by Johan Maurits of Nassau-Siegen to Frederick III (1609–1670 | King of Denmark).

Copy 2: Shelf mark: Fol. N. Hist. 13970. Binding: Eighteenth-century parchment, spine with red title label. Coloration: Black-and-white. Complete: Yes.

Finland

Helsinki, National Library of Finland

Collection: A.E. Nordenskiöld Collection. Shelf mark: N. 509. Binding: Contemporary parchment. Coloration: Black-and-white. Complete:

Yes. Provenance: Bookplate: [Removed]. Manuscript annotation: "…
vendu 20-30 f." From the library of Adolf Erik Nordenskiöld (1832–1901 |
Finnish polar explorer). Acquired as a whole by the former Helsinki
University Library in 1902.

France

Aix-en-Provence, Bibliothèque Méjanes

Copy 1: Shelf mark: In Fol. 0203.

Copy 2: Shelf mark: In Fol. 0212.

Avignon, Bibliothèque Municipale

Shelf mark: Fol. 1798.

Besançon, Bibliothèque Municipale

Shelf mark: 10926. Binding: Contemporary calf, gilded and decorated
with filets.
  Provenance: Ex libris [not identified]. Manuscript annotations.

Boulogne-sur-Mer, Bibliothèque Municipale

Shelf mark: C 6177.

Carpentras, Bibliothèque Inguimbertine

Collection: Collection d'Inguimbert. Shelf mark: E 1794. Provenance:
Donated to the library by Joseph-Dominique Malachie d'Inguimbert
(1683–1757 | French prelate and librarian, bishop of the Diocese of
Carpentras).

Chalon-sur-Saône, Bibliothèque Municipale

Shelf mark: in-2 245.

Bordeaux, Bibliothèque Municipale de Bordeaux

Shelf mark: S 824. Binding: Nineteenth-century half-leather. Provenance:
Library stamp: l'Académie Royale de Bordeaux. Library stamp:
Bibliothèque de la Ville de Bordeaux.

Dijon, Bibliothèque Municipale

Shelf mark: 11023.

Grenoble, Bibliothèque d'Etude et du Patrimoine

Shelf mark: A.712.

Le Mans, Médiathèque Louis Aragon

Shelf mark: SA F* 1423.

Lyon, Bibliothèque Municipale

Copy 1: Shelf mark: Rés 22720. Binding: Contemporary parchment, gilded coat of arms on front cover. Coloration: Black-and-white. Complete: Yes. Provenance: Coat of arms is of M.A. Mazanot. Manuscript annotation on folium *2: Jésuites de Lyon 1660. Library stamp: Colleg. Lugdun. Library stamp: Bibliothèque de la ville de Lyon.

Copy 2: Shelf mark: Rés 30656. Binding: Eighteenth-century mottled calf, marbled endpapers. Coloration: Black-and-white. Complete: Yes. Provenance: Armorial bookplate: Petrus Adamoli [Pierre Adamoli] (1707–1769 | French bibliophile), dated 1733. Library stamp: Acad Scient. Litt. et Art. Lugd. [L'Académie des sciences, belles-lettres et arts de Lyon].

Marseille, Bibliothèque Municipale à Vocation Régionale

Copy 1: Collection: Bibliothèque des Bernardines. Shelf mark: Xb3884.

Copy 2: Shelf mark: Xb3885.

Montpellier, Bibliothèque Interuniversitaire

Shelf mark: Hist. Médecine, Da 33 in-fol.

Nancy, Bibliothèque Stanislas

Shelf mark: 101642. Binding: Parchment.

Nantes, Bibliothèque Municipale

Copy 1: Shelf mark: 13289A. Provenance: Ex libris: Jos. Arnoult.

Copy 2: Shelf mark: 13289B.

Nice, Bibliothèque Patrimoniale et d'étude Romain Gary

Shelf mark: XVII-4049. Binding: Contemporary natural calf.

Paris, Bibliothèque Interuniversitaire de Santé

Copy 1: Location: Pôle médecine-odontologie. Binding: Blind stamped pigskin over wooden boards. Shelf mark: 903-1. Coloration: Black-and-white. Complete: Yes. Provenance: Manuscript annotation on title page: Collegii Soctis Jesu Coloniae 1668.

Copy 2: Location: Pharmacie. Shelf mark: RES 75. Binding: Modern binding to replace a defective blind-tooled binding. Coloration: Black-and-white. Complete: no, lacks folium 2F4.

Paris, Bibliothèque Interuniversitaire de la Sorbonne

Shelf mark: SND 3= 13. Binding: Contemporary mottled calf. Coloration: Black-and-white. Complete: Yes. Provenance: From the Bibliothèque du Prytanée. Library stamp: Bibliothèque de l'Université de Paris. Acquired by the library from the legacy of Jean-Gabriel Petit de Montempuis (1676(?)–1763 | French philosophy professor).

Paris, Bibliothèque Mazarine

Shelf mark: 2° 4029. Binding: Contemporary natural calf. Coloration: Black-and-white. Complete: Yes. Provenance: From the library of Cardinal Jules Raymond Mazarine (1602–1661 | Italian cardinal, diplomat, and politician).

Paris, Bibliothèque Nationale

Copy 1: Collection: Réserve des livres rares. Shelf mark: S-851. Binding: Contemporary Dutch vellum with gold tooling. Coloration: Colored. Complete: Yes.

Copy 2: Collection: Réserve des livres rares. Shelf mark: RES-S-258. Binding: Contemporary natural calf with coat of arms of Gaston d'Orléans. Coloration: Black-and-white. Complete: Yes. Provenance: From the collection of Gaston d'Orléans (1608–1660 | Duke of Orléans). Included in the collections of the Bibliothèque Nationale through legacy of the Royal collections.

Copy 3: Collection: Bibliothèque de l'Arsenal. Shelf mark: FOL-S-437. Binding: Contemporary brown calf. Coloration: Black-and-white. Complete: Yes. Provenance: Manuscript annotation: J.-B. Chomel D.M.P. [Pierre-Jean-Baptiste Chomel] (1671–1740 | botanist). From the Bibliothèque de La Vallière, sold in 1786 (catalogue Nyon, no. 5108).

Copy 4: Collection: Bibliothèque de l'Arsenal. Shelf mark: FOL-S-438. Binding: Contemporary marbled brown calf. Coloration: Black-and-white. Complete: Yes. Provenance: Manuscript annotation: Carolus Feron.

Copy 5: Collection: Bibliothèque de l'Arsenal. Shelf mark: FOL-S-439. Binding: Contemporary natural calf, with ex dono of Léonard Tardi (d.1671 | auditor at the Chambres des Comptes). Coloration: Black-and-white. Complete: no, lacks title page. Provenance: From the library of Léonard Tardi, donated in 1671 to the Bibliothèque des Grands Augustins.

Paris, Collège France, Bibliothèque Générale

Collection: Bibliothèque patrimoniale. Shelf mark: XV Fol 25. Coloration: Black-and-white. Complete: Yes.

Paris, Conservatoire National des Arts et Métiers

Shelf mark: Fol. Y 13. Binding: Contemporary calf over wooden boards, blind stamped oval medallion with the coat of arms of Vincenty de Vischer on front cover, decorated with geometric and floral motifs, traces of clasps. Coloration: Black-and-white. Complete: Yes. Provenance: Ex dono Vincenty de Vischer (mid-seventeenth century | Abbey of Grimbergen).

Paris, Institut de France

Collection: Collection Benjamin Delessert. Shelf mark: Fol DM 192. Binding: Contemporary natural calf. Coloration: Black-and-white. Complete: Yes. Provenance: Bookplate [not identified]. Acquired by the library in 1869.

Paris, Muséum National d'Histoire Naturelle

Copy 1: Collection: Bibliothèque centrale. Shelf mark: Fol Bn 91.

Copy 2: Collection: Bibliothèque centrale. Shelf mark: 24 698.

Périgueux, Médiathèque Pierre Fanlac

Shelf mark: SA-III-1 B 803. Provenance: Manuscript annotation: Abbatia Beata Maria de Cancellata catal.inscript. From the Abbaye Notre-Dame de Chancelade.

Rennes, Les Champs Libres

Shelf mark: 3633.
   Rochefort, Bibliothèque de Rochefort Musée national de la Marine
   Shelf mark: 44 H.

Strasbourg, Bibliothèques Universitaires de Strasbourg

Location: Bibliothèque Huet-Weiller. Collection: Sciences Magasin Sous-sol. Shelf mark: HR 8. Coloration: Black-and-white. Complete: Yes. Provenance: Stamp on title page: Pharmaceutisches Institut Universität Strassburg.

Germany

Augsburg, Staats- und Stadtbibliothek

Shelf mark: 2° Nat 109. Binding: Seventeenth-century parchment. Coloration: Black-and-white. Complete: Yes. Provenance: Manuscript annotation on title page: Lukas Schröck (1646–1730 | German medical doctor). Bequeathed to the Stadtbibliothek Augsburg in 1730. Note: Copy bound with *De Indiæ utriusque re naturali et medica libri qvatvordecim* (Amstelædami, apud L. and D. Elzevirios, 1658).

Bad Arolsen, Fürstlich Waldecksche Hofbibliothek Arolsen

Shelf mark: III 1a 1. Binding: Contemporary parchment. Coloration: Black-and-white. Complete: Yes. Provenance: Manuscript annotation: [illegible] 4/14 Nov. 1649 J. Dillenburg [Possibly from the family of Johannes Wilhelmus Dillenburg (1646–1696 | medical doctor)]. From the Schaumburgische Bibliothek of Schloss Schaumburg bei Diez an der Lahn. Acquired through heritage and purchase by George Victor (1831–1893 | Prince of Waldeck and Pyrmont) and subsequently entered in the Fürstlich Waldecksche Hofbibliothek.

Bamberg, Staatsbibliothek

Shelf mark: 22/H.n.f.16.

Berlin, Staatsbibliothek

Copy 1: Location: Unter den Linden, Abteilung Historische Drucke. Shelf mark: 2" Lh 11450. Coloration: Colored. Complete: Yes. Provenance: Manuscript annotation: Ottho L.B. à Schwerin (1645–1705 | diplomat). Manuscript annotation: Karl Asmund Rudolphi (1771–1832 | Swedish naturalist). Note: According to Martin Hinrich Lichtenstein (1780–1857 | German naturalist) this copy, brought in by Rudolphi, was owned and annotated by Johan Maurits of Nassau-Siegen. If so, this copy might have been part of the sale of the original drawings by Albert Eckhout to Friedrich Wilhelm, Elector of Brandenburg in 1652.

Copy 2: Location: Unter den Linden, Abteilung Historische Drucke. Shelf mark: Bibl. Diez fol. 196. Coloration: Black-and-white. Complete: Yes. Provenance: Manuscript annotation: Societatis Scient. [with old shelf mark] im Buchdeckel: Fol: Phys. n 42 T 1.

Copy 3: Shelf mark: 2" Lh 11450. Note: Not available for consultation, this copy was probably lost in the Second World War.

Bonn, Universitäts- und Landesbibliothek

Copy 1: Shelf mark: Qa 2' 109. Binding: Contemporary parchment. Coloration: Black-and-white. Complete: Yes. Provenance: One of the copies in Bonn is from the collection of Karl Wilhelm Nose (1753–1835 | German medical doctor), donated in two parts in 1819 and 1827. It cannot be verified if this is copy 1 or 2.

Copy 2: Shelf mark: Qa 2' 109 #1. Binding: Restored with a new binding in 1992, originally parchment or half-parchment. Coloration: Black-and-white. Complete: No, lacks page 293–294. Provenance: Manuscript annotation on title page: Sum ex Bibliotheca Immanuelis Brigelii [probably Emanuel Brigel? (fl. 1663 | German naturalist)]. Crossed out manuscript annotation on title page: [illegible]. Bookseller's ticket: Antiquarisches Lager der F.F. Autenriethschen Buchhandlung in Stuttgart.

Erlangen, Universitätsbibliothek Erlangen-Nürnberg

Shelf mark: H00/MED-III 4. Binding: Contemporary parchment, with a supralibros of Universitas Altorfina on front and back cover. Coloration: Black-and-white. Complete: Yes. Provenance: From the library of the University of Altdorf. After the university was closed in 1809, the collection was transferred to the University Library of Erlangen in 1818.

Frankfurt am Main, Goethe-Universität, Universitätsbibliothek
Johann Christian Senckenberg

Shelf mark: 2° 8. Binding: Parchment. Coloration: Black-and-white. Complete: Yes. Provenance: Library stamp on title page: Senckenbergische Bibliothek Frankfurt am Main. Stamp: Mr. Carl von Heyden (1793–1866 | German politician and entomologist, co-founder of the Senckenbergischen Naturforschenden Gesellschaft). Note: Manuscript marginalia, probably by Carl von Heyden.

Freiberg, TU Bergakademie Freiberg, Universitätsbibliothek

Shelf mark: V 343 2.

Göttingen, Niedersächsische Staats- und Universitätsbibliothek

Shelf mark: 2 H NAT III, 5806. Coloration: Black-and-white. Complete: Yes. Provenance: Library stamp: Ex bibliotheca Regia Acad. Georgiae Aug. [University Library Göttingen].

Halle, Universitäts- und Landesbibliothek Sachsen-Anhalt

Shelf mark: Oc 6177, 2°. Binding: Parchment (defective). Coloration: Black-and-white. Complete: Yes. Provenance: Bibliotheca Leucorea. Universitätsbibliothek Wittenberg.

Hannover, Gottfried Wilhelm Leibniz Bibliothek

Shelf mark: Gp-A 10051. Binding: Contemporary parchment. Coloration: Black-and-white. Complete: Yes. Provenance: Manuscript annotation: Martini Fogeli Hamburg [Martinus Fogelius] (1634–1675 | German medical doctor). Collection of Fogelius was acquired as a whole by Gottfried Wilhelm Leibniz (1646–1716 | German philosopher) for Johann Friedrich von Braunschweig-Lüneburg (1625–1679 | duke of Brunswick-Lüneburg) in 1678.

Heidelberg, Universitätsbibliothek

Shelf mark: O 5300 Gross RES. Binding: Leather over paper boards, gilded adornment lines, gilded emblem of Iac. Aug. Thvanvs on front cover and back cover. Coloration: Black-and-white. Complete: Yes.

Jena, Thüringer Universitäts- und Landesbibliothek

Copy 1: Collection: Herbarium Haussknecht. Shelf mark: M2PISO. Binding: Contemporary blind-tooled parchment (defective). Coloration: Black-and-white. Complete: No, lacks title page, page 47–48 defective, replaced in manuscript.

Copy 2: Shelf mark: 2 Hist.nat.VI,11. Binding: Contemporary parchment, edges gilt. Coloration: Black-and-white. Complete: Yes.

Copy 3: Shelf mark: 2 Hist.nat.VI,12. Note: Copy is missing.

Leipzig, Universitätsbibliothek

Shelf mark: Allg.N.W.69. Binding: Contemporary parchment. Coloration: Black-and-white. Complete: Yes. Provenance: Manuscript annotation: K. Leineker D. Manuscript annotation: ſ 5, 1707. Manuscript annotation: Emi in Auct. Wittweriana [Auction of the collection of Philipp Ludwig Wittwer (1752–1792 | medical doctor) in 1794].

Mannheim, Universitätsbibliothek

Shelf mark: Sch 106/332. Binding: Contemporary parchment. Coloration: Black-and-white. Complete: Yes. Provenance: Manuscript annotation on title page: [illegible name] 1738. Library stamp: Bibliothek Desbillons Mannheim. From the library of François-Joseph Terrasse Desbillons (1711–1789 | French Jesuit and author), transferred to Mannheim in 1764 after the suppression of the Jesuits in France.

Müncheberg, Senckenberg Deutsches Entomologisches Institut

Location: Entomologische Bibliothek. Shelf mark: B 966. Binding: Later half-cloth. Coloration: Black-and-white. Complete: Yes.

Munich, Bayerische Staatsbibliothek

Location: BSB/Handschriftenabt. Magazin. Shelf mark: Rar. 2208. Binding: Contemporary parchment over paper boards. Coloration: Black-and-white. Complete: Yes. Provenance: Library stamp: Ex bibliotheca

Academiae Julia Carolina Helmstadii [Library of the University of Helmstad, dissolved in 1810].

Munich, Bibliothek der Ludwig-Maximilians-Universität

Copy 1: Location: Zentralbibliothek. Shelf mark: 0014/W 2 H.nat. 15. Binding: Contemporary pigskin over wooden boards. Coloration: Black-and-white. Complete: Yes. Provenance: Bookplate: Christoph Jacob Trew (1695–1769 | medical doctor, botanist). The collection of Trew was originally incorporated by the University of Altdorf. After the university was closed in 1809, the collection was transferred to the University Library of Erlangen in 1818. This copy was given as a doublet to the University of Landshut, a predecessor of Ludwig-Maximilians University.

Copy 2: Location: Georgianum. Shelf mark: 0017/2 Hist.prof. 65. Binding: Contemporary leather. Coloration: Black-and-white. Complete: Yes.

Oldenburg, Landesbibliothek

Shelf mark: NW II 1 325. Binding: Contemporary leather, gilded stamp of an angel reading a book, with the letters B.E.R.P. on front and back cover. Coloration: Colored. Complete: Yes. Provenance: Acquired in the eighteenth century by Georg Friedrich Brandes (1709–1791 | German jurist, art collector, and bibliophile). Bought by Herzog Peter Friedrich Ludwig (1755–1829 | Regent of the Duchy of Oldenburg) in 1790. Subsequently included in the Landesbibliothek Oldenburg.

Rostock, Universitätsbibliothek

Shelf mark: 28-RAR:Na-8. Binding: Half-leather with marbled boards, spine decorated with gold. Coloration: Black-and-white. Complete: Yes.

Stuttgart, Württembergische Landesbibliothek

Shelf mark: Nat.G.fol.436. Binding: Contemporary brown calf, gold embossing on front cover, edges gilt, marbled endpapers. Coloration: Black-and-white. Complete: Yes.

Tübingen, Universitätsbibliothek

Copy 1: Shelf mark: Fo XXVIII 1.2. Binding: Contemporary parchment. Coloration: Black-and-white. Complete: Yes. Provenance: On paste-down: Ad Bibliothec. Aulic. Elvacensem [Library of Ellwangen Abbey]. Acquired by the library in mid-nineteenth century.

Copy 2: Shelf mark: Bg 15.2. Binding: Contemporary parchment. Coloration: Black-and-white. Complete: Yes. Provenance: Title page with manuscript dedication to Theodorus Quingerus[?].

Weimar, Herzogin Anna Amalia Bibliothek

Copy 1: Shelf mark: Scha BS 1 C 05098 (1). Provenance: Manuscript annotation: Konrad Samuel Schurzfleisch (1641–1708 | German historian, librarian) 1707. Note: Not available for consultation, this copy was damaged in the fire of 2004.

Copy 2: Shelf mark: 19 C 12817. Binding: Contemporary pigskin over wooden boards. Coloration: Black-and-white. Complete: Yes. Provenance: Bookplate: Gaming 1653. Acquired through an auction at Reiss & Sohn, Königstein im Taunus in 2007.

Wolfenbüttel, Herzog-August-Bibliothek

Shelf mark: A: 6.6 Phys. 2° (1). Note: Mentioned in the Bücherradkatalog on page 1001.

Ireland

Dublin, National Library of Ireland

Shelf mark: LBR 591981. Binding: Rebound at the National Library of Ireland on 6 February 1896. Coloration: Black-and-white. Complete: Yes. Provenance: Library stamp: National Library of Ireland.

Dublin, Trinity College Library

Collection: Fagel collection. Shelf mark: Fag. M.1.6. Binding: eighteenth-century sprinkled calf, with edges of front and back boards rolled in gold; spine, with seven raised bands, stamped and rolled in gold, with black calf shelf mark label, lacking matching title-label; marbled edges and endpapers. Coloration: Colored. Complete: Yes. Provenance: Purchased from Hendrik Fagel "the younger" (1765–1838 | Dutch greffier of the States General 1790–1795) for Trinity College as part of the Fagel Collection in 1802. Previously owned by other members of the Fagel family.

Italy

Bassano del Grappa, Biblioteca civica

Shelf mark: REC 5.D.6. Binding: Contemporary parchment. Coloration: Black-and-white. Complete: Yes. Provenance: Bookseller's ticket: Presso

Pietro Agnelli Librario, e Stampatore in Milano Santa Margherita. Donated to the library by Giambattista Brocchi (1772–1826 | Italian naturalist, mineralogist, and geologist).

Bologna, Biblioteca comunale dell'Archiginnasio

Shelf mark: 527, 32. I.00 00017. Binding: Half-leather.

Bologna, Università di Bologna

Copy 1: Location: Biblioteca del Dipartimento di scienze biologiche, geologiche e ambientali. Shelf mark: BOV 1032, Irnerio Bertoloni 008 A 009. Provenance: Library stamp: Orto botanico di Bologna. Manuscript annotation: F. Giovannini (custodian of the garden).

Copy 2: Location: Biblioteca del Dipartimento di scienze biologiche, geologiche e ambientali. Shelf mark: BOV 2096, Irnerio Bertoloni 004 A 003. Provenance: Manuscript annotation: Dedication by Johannes de Laet (1581–1649 | Dutch geographer, editor of *Historia Natvralis Brasiliæ*) to Cassiano dal Pozzo (1588–1657 | Italian scholar, book and arts collector). Library stamp: Biblioteca Albani.

Catania, Biblioteca Regionale Universitaria

Shelf mark: LC 4.219. Binding: Parchment (defective).

Faenza, Biblioteca Comunale Manfrediana

Shelf mark: H 003 007 011. Provenance: Lodovico Caldesi (1821–1884 | Italian botanist). Caldesi donated his library as a whole to the Biblioteca Comunale Manfrediana.

Fermo, Biblioteca Civica Romolo Spezioli

Shelf mark: 1 O 9/9113. Binding: Contemporary parchment, spine with five raised bands, manuscript title in second compartment. Coloration: Black-and-white. Complete: Yes. Provenance: Old shelf mark: F 7/14; 209. Manuscript annotation on paste-down: Ex libris Romuli Spetioli Firmani. From the collection of Romolo Spezioli (1642–1723 | Italian medical doctor and the personal physician of Queen Christina of Sweden, Cardinal Decio Azzolino, and of Pope Alexander VIII). Donated by Spezioli to the library in 1705, or through legacy in 1723 as part of his collection of 12.000 volumes. Note: Engraved title page mutilated with ink to cover the naked parts of both figures.

Ferrara, Biblioteca Comunale Ariostea

Shelf mark: N 9.11.13. Binding: Parchment. Provenance: Manuscript annotation: Libreria de Capuccini di Ferrara.

Florence, Biblioteca Nazionale Centrale di Firenze

Copy 1: Shelf mark: MAGL.1._.90.

Copy 2: Shelf mark: PALAT.2.6.6.11.

Genoa, Biblioteca Universitaria di Genova

Shelf mark: Rari XIV 34. Binding: Contemporary parchment. Coloration: Black-and-white. Complete: Yes. Provenance: Held by Regia Università di Genova since 1853.

Milan, Biblioteca Nazionale Braidense

Shelf mark: C. 17. 09607. Binding: Parchment.

Naples, Biblioteca Nazionale Vittorio Emanuele III

Copy 1: Collection: Fondo Farnese. Shelf mark: Sala Farnese XXIX G 13. Binding: Contemporary parchment. Coloration: Black-and-white. Complete: Yes. Provenance: Manuscript annotation on title page: [Illegible, probably of Jesuit origin]. Probably acquired at the end of the eighteenth century with the suppression of religious orders.

Copy 2: Collection: Fondo Doria. Shelf mark: F.Doria IV 330. Binding: Contemporary parchment. Coloration: Black-and-white. Complete: Yes. Provenance: From the legacy of Gino Doria (1888–1975 | Italian journalist and historian).

Padua, Biblioteca civica

Shelf mark: E. 1384.

Padua, Biblioteca Universitaria di Padova

Shelf mark: 101.A.10. Binding: Cardboard. Coloration: Black-and-white. Complete: Yes. Provenance: Stamp of the Republic of Venice era; entered the library before 1797.

Padua, Università degli Studi di Padova

Collection: Biblioteca dell'orto botanico. Shelf mark: APL.291. Binding: Eighteenth-century brown leather, gold stamping on back cover, marbled endpapers. Coloration: Black-and-white. Complete: Yes. Provenance: From the library of Giovanni Marsili (1727–1795 | Italian botanist, prefect of the botanical garden of Padua from 1760 to 1794); present in the manuscript catalogue of the collection of Marsili as number 1817. Purchased as a whole by Antonio Bonato (1753–1836 | Italian botanist, prefect of the Botanical garden of Padua from 1795 to 1835). Bequeathed as a whole to the University in 1835.

Pavia, Biblioteca della Scienza e della Tecnica

Collection: Fondo Santo Garovaglio. Shelf mark: Orto Botanico SALA.28.D.7 20. Binding: Parchment. Coloration: Black-and-white. Complete: Yes. Provenance: Purchased by Santo Garovaglio (1805–1882 | Italian botanist, prefect of the Botanical garden of Pavia).

Rome, Biblioteca Casanatense

Shelf mark: O I 23. Binding: Contemporary blind-tooled parchment over boards. Coloration: Black-and-white. Complete: Yes. Provenance: Not mentioned in the index of printed books of Girolamo Casanate that was compiled around 1682–1687 but listed in the nineteenth-century manuscript catalogue. Probably acquired in the eighteenth or early nineteenth century.

Rome, Biblioteca Nazionale Centrale di Roma

Collection: Fondo Antico. Shelf mark: 68. 3.H.5. Binding: Rebound in leather by A. Lombardi in 1972. Coloration: Black-and-white. Complete: Yes. Provenance: Gesuiti Collegio Romano: Preposto generale. Library stamp: Biblioteca Nazionale.

Turin, Biblioteca Nazionale Universitaria

Shelf mark: 1-659871. Binding: Contemporary leather, decorated in gold.

Venice, Biblioteca Nazionale Marciana

Shelf mark: D 200D 018. Binding: Eighteenth-century leather, decorated in gold, marbled edges. Coloration: Black-and-white. Complete: Yes. Provenance: Book plate on verso engraved title page: Ex libris Marciano

(Bragaglia 512), made in 1722 on behalf of librarian Girolamo Venier (1650–1735 | Italian composer and librarian of Marciana from 1709 until 1735) to mark parts of the collection. Note: Old shelf mark: XX.3.

Venice, Ca' Foscari Fondo Storico

Shelf mark: BG 11.B.10.

Venice, Museo di Storia Naturale di Venezia

Shelf mark: C 24 Q MSNVE. Binding: Eighteenth-century leather. Coloration: Black-and-white. Complete: Yes. Provenance: Contarini [Italian noble family, one of the founding families of Venice]. Bookplate: Museo Civico Correr; placement D 9.

Vicenza, Biblioteca Civica Bertoliana

Shelf mark: X 023 008 022 MAGAZZINO.

Mexico

Biblioteca Nacional de México

Collection: Fondo de Origen. Shelf mark: 94-43944.

The Netherlands

Amsterdam, Bibliotheek Nederlands Tijdschrift voor Geneeskunde

Shelf mark: 24 G 10. Binding: Contemporary brown leather, decorated spine with six raised bands. Coloration: Black-and-white. Complete: Yes. Provenance: Bookplate: Nederlandsch Tydschrift voor Geneeskunde.

Amsterdam, Scheepvaartmuseum

Shelf mark: Me-0506. Binding: Eighteenth-century marbled boards. Coloration: Black-and-white. Complete: Yes. Provenance: From the collection of Anton Mensing (1866–1936 | Dutch art dealer and collector). Originally on loan to the Scheepvaartmuseum from Mensing. Permanently acquired by the museum at the Mensing auction hosted by Sotheby's in 1936. Reference: *The Mensing Library Catalogue of the very valuable and important library … first portion*, Lot. no. 462. London: Sotheby, 1936.

Amsterdam, Universiteitsbibliotheek

Copy 1: Location: Artis Bibliotheek. Shelf mark: AB 090:18. Binding: Contemporary parchment with gold tooling. Coloration: Black-and-white. Complete: Yes. Provenance: Manuscript annotation: Joh[annes] Bon (ca. 1720–1802 | medical doctor). Ex libris: Cornelis Henricus à Roy (1750–1833 | medical doctor). Library stamp: Koninklijk Zoölogisch Genootschap Natura Artis Magistra. Note: The man and woman on the engraved title page are covered with a leaf.

Copy 2: Location: Allard Pierson. Shelf mark: OTM: KF 61-4353. Binding: Contemporary parchment, gilded text on front cover: Hortus Medicus Amstelodamensis. Coloration: Black-and-white. Complete: Yes. Provenance: From the collections of the Hortus Botanicus in Amsterdam. Transferred to the City Library, which would later become the University Library, no later than 1861.

Amsterdam, Vrije Universiteit

Shelf mark: XQ.05049. Binding: Contemporary marbled brown leather, gilded coat of arms on front cover. Coloration: Black-and-white. Complete: Yes.

Groningen, Rijksuniversiteit

Shelf mark: MF1. Binding: Contemporary brown leather, panel design with gold tooling on front cover, spine with five raised bands and gilded floral decoration. Coloration: Black-and-white. Complete: Yes. Provenance: Manuscript annotation (Latin): Donated to Groningen University Library by "the consul and senators" of the States of Groningen in 1668. Library stamp on title page: Groninganae Bibliotheca Academiae.

Haarlem, Teyler's Museum

Copy 1: Shelf mark: 5G 5. Binding: Contemporary brown leather, panel design with blind tooling on front cover, spine with five raised bands. Coloration: Black-and-white. Complete: Yes. Provenance: Library stamp: Bibliothèque Musée Teyler Harlem.

Copy 2: Shelf mark: 113 AB21300. Binding: Eighteenth-century green morocco, rebacked gilt decorated spine with floral motifs and six raised bands, all edges gilt, marbled endpapers. Coloration: Black-and-white. Complete: Yes. Provenance: Bookplate: Ex Bibliotheca Com: Thomae Cajetani de Węgry Węgierski [Tomasz Kajetan Węgierski] (1756–1787 | Polish poet). Booksellers' plate: Rey et Gravier, quai des Augustins, no. 55, Paris. Manuscript annotation: Dr. Walter Channing (1786–1876 | American physician and professor of medicine), 11 April 1840; sent as a gift to Amos Binney, includes a two-page handwritten letter to accompany the donation. Manuscript annotation: Amos Binney (1803–1847 | American physician and malacologist). Bookplate: Boston Society of Natural History, from the library of Dr. Amos Binney, deposited by Mrs. M.A. Binney. Manuscript annotation: Col. Büch. Acquired by Boudewijn Büch (1948–2002 | Dutch author and book collector) from Kistner and Ackerman (Nürnberg/München | Antiquarian bookdealers) in January 1993.

The Hague, Gemeentearchief

Copy 1: Shelf mark: Bibl. de Cocq fo 42. Binding: Contemporary parchment. Coloration: Black-and-white. Complete: Yes.

Copy 2: Shelf mark: Bibl. de Cocq fo 49 (CPh). Binding: Contemporary parchment. Coloration: Black-and-white. Complete: Yes. Provenance: Manuscript annotation (Latin): M[ichiel] Boudewijns (1750–1833 | medical doctor | Antwerp).

The Hague, Koninklijke Bibliotheek

Shelf mark: 36 C 3. Binding: Contemporary parchment, gilded capital Y on front and back cover, spine restored. Coloration: Black-and-white. Complete: Yes. Provenance: From the library of the Friars house, affiliated with the Latin School in Delft.

Leeuwarden, Tresoar

Shelf mark: 700 Ntk fol. Binding: Contemporary brown calf over wooden boards, panel design with blind stamped coat of arms, two metal claps. Coloration: Black-and-white. Complete: Yes. Provenance: Added folium with printed text (Latin) on both sides: Donated by Willem Frederik van Nassau-Dietz (1613–1664 | Stadtholder of Friesland, Groningen and Drenthe) to Franeker University Library in 1649. Library stamp on title page: P.B.v.F. [Provinciale Bibliotheek van Friesland].

Leiden, Naturalis Biodiversity Center

Copy 1: Shelf mark: RBR D00545. Binding: Contemporary parchment with gold embossing, all edged gilt, spine with six raised bands. Coloration: Colored. Complete: Yes. Provenance: Ex libris: Ferdinand

Casper Koch, Rotterdam (1873–1957 | Dutch jurist and book collector). Blind stamp: Rijksmuseum van Natuurlijke Historie. Purchased by the Museum in 1974.

Copy 2: Shelf mark: RBR Holt 00563. Binding: Rebacked brown leather, black label on spine, five raised bands. Coloration: Black-and-white. Complete: Yes. Provenance: Manuscript annotation on folium A1: Tho[mas More] Molyneux (1661–1733 | Irish physician). Manuscript annotation on title page: [Illegible] 1791. Library stamp: [Faded]. From the collection of Lipke Bijdeley Holthuis (1921–2008 | Dutch carcinologist and book collector). Bequeathed as a whole to the museum in 2008.

Copy 3: Shelf mark: mus-nev 61651. Binding: Contemporary brown leather, spine with six raised bands, red title-label and decoration in gold. Coloration: Black-and-white. Complete: Yes. Provenance: Stamp: [Hendrick] Hartogh Heijs van de Lier, Delft (1821–1870 | Dutch entomologist). Library stamp: Nederlandse Entomologische Vereniging. Donated by the widow of Hartogh Heijs van de Lier to the Nederlandse Entomologische Vereniging in 1870.

Leiden, Universiteitsbibliotheek

Copy 1: Collection: Bibliotheca Thysiana. Shelf mark: THYSIA 2274. Binding: Contemporary parchment, blind tooling, spine with six raised bands. Coloration: Black-and-white. Complete: Yes. Provenance: Manuscript annotation on title page: "sum J. Thijs." Johannes Thijs (1622–1653 | Dutch book collector).

Copy 2: Shelf mark: 1407 B 3. Binding: Contemporary brown leather, gold tooling on front cover. Coloration: Colored. Complete: Yes. Provenance: Library stamp on title page and edges: Acad Lvgd Bat [Leiden University].

Copy 3: Collection: KITLV Shelf mark: M 3t 54. Binding: Blind tooled contemporary parchment, spine with six raised bands. Coloration: Black-and-white. Complete: Yes. Provenance: Library stamp on title page: Indisch Genootschap. From the collection of the Koninklijk Instituut voor Taal-, Land- en Volkenkunde, transferred to Leiden University Library in 2014.

Copy 4: Shelf mark: VDSAND 236 A 12. Binding: Contemporary brown leather, gilded coat-of-arms on front cover: two shamrocks and two fleur-de-lis in a cartouche. Coloration: Black-and-white. Complete: Yes. Provenance: Faded text (Danish) in old hand on front cover and on pastedown: [illegible]. Bookplate: Bibliotheca Pharmacia Van de Sande. Acquired by J.M.H. (Jaap) van de Sande (1916–2001 | Dutch pharmacist) from Norlis Antikvariat, Oslo.

Utrecht, Universiteitsbibliotheek

Shelf mark: R fol 13. Binding: Contemporary blind-stamped parchment. Coloration: Black-and-white. Complete: No, lacks pages 259–262. Provenance: Library stamp: Academia Rheno-Traiectina [Utrecht University]. Note: Annotated throughout in old hand, contains a handwritten index.

Wageningen, University & Research, Library

Shelf mark: Forum Library, R333A03. Binding: Contemporary leather over wooden plates, gilded decoration on back cover. Binding restored in 1984. Coloration: Black-and-white. Complete: Yes. Provenance: Library stamp: Bibliotheek der Landbouwhoogeschool [used from 1918 to 1990].

Poland

Gdansk, Polska Akademia Nauk Biblioteka Gdańska

Shelf mark: Uph. f. 1483. Binding: Parchment. Coloration: Black-and-white. Complete: Yes. Provenance: Monogrammed: A.K.D. Ex Bibliotheca Uhliana [Uhl Tobias Christian] (1743–1795 | Polish clerk). Johann Uphagen (1731–1802 | Polish bibliophile).

Kraków, Biblioteka Jagiellońska

Copy 1: Shelf mark: BJ St. Dr. Przyr. 258. Binding: Green-brown marbled paper boards. Coloration: Black-and-white. Complete: Yes. Provenance: "Ex Libris Valen[...] Francisci Gientu [= Walenty Franciszek?] Consulis Tarno [viensis].

Copy 2: Collection: Księgozbiór Kamedułów. Shelf mark: BJ Cam. L. XV. 2. Binding: Parchment. Complete: Lacks frontispiece. Provenance: From the library of the Camaldolese Priory in Bielany, Kraków.

Portugal

Lisbon, Biblioteca Nacional de Portugal

Copy 1: Collection: Duarte de Sousa. Shelf mark: D.S. XVII – 77. Binding: Eighteenth-century red morocco, gold tooling, coat of arms, and floral decoration on front and back cover, spine with five raised bands. Book restorer's ticket on pastedown: Frederico D'Almeida, Encadernador, Lisboa. Coloration: Black-and-white. Complete: Yes. Provenance: António Alberto Marinho Duarte de Sousa (1896–1950 | Portuguese bibliophile).

Copy 2: Shelf mark: ELZ. 345. Binding: Parchment with traces of old ribbons. Coloration: Black-and-white. Complete: Yes. Note: Contains an additional folium with manuscript annotations in English, possibly nineteenth century, attached to pastedown, with a synthesis of the history of "William Piso."

Copy 3: Shelf mark: ELZ. 348. Binding: Paper boards. Coloration: Black-and-white. Complete: Yes. Provenance: Stamp on last page: D. Franc. Manuel. From the library of D. Francisco de Melo Manuel (Cabrinha) (1773–1851 | Portuguese bibliophile).

Copy 4: Shelf mark: ELZ. 349. Binding: Blind stamped parchment. Coloration: Black-and-white. Complete: Yes. Provenance: Stamp on last page: D. Franc. Manuel. From the library of D. Francisco de Melo Manuel (Cabrinha) (1773–1851 | Portuguese bibliophile).

Copy 5: Shelf mark: ELZ. 392. Binding: Brown leather. Coloration: Black-and-white. Complete: Yes.

Coimbra, UC Biblioteca Geral

Copy 1: Collection: Biblioteca de São Pedro. Shelf mark: S.P.-O-7-3. Binding: Sheepskin over paper boards, spine with six raised bands. Coloration: Black-and-white. Complete: Yes. Provenance: Stamp and manuscript annotation: Real Colégio de São Pedro de Coimbra.

Copy 2: Collection: Biblioteca Joanina. Shelf mark: 4-10-38-6 c.2. Binding: Parchment. Coloration: Black-and-white. Complete: Yes.

Russia

Moscow, Russian State Library

Copy 1: Shelf mark: МК [Музей книги] Amsterdam Elzevier 1648 2°. XVII-1223.

Copy 2: Shelf mark: МК [Музей книги] Amsterdam Elzevier 1648 2°. MK VIII-11695.

Copy 3: Shelf mark: МК [Музей книги] Amsterdam Elzevier 1648 2°. MK VIII-11696.

Saint Petersburg, Russian Academy of Sciences Library

Binding: Restored binding. Coloration: Black-and-white. Complete: Yes. Provenance: Library stamp on title page, indicating that this copy was probably acquired before 1741–1744.

Spain

Barcelona, Universitat de Barcelona, CRAI Biblioteca de Reserva

Shelf mark: 07 C-195/1/13. Binding: Contemporary limp vellum, manuscript capital P on spine, paper restored in 1999. Coloration: Black-and-white. Complete: Yes. Provenance: From the collection of Josep Jeroni Besora (d. 1665 | Catalan cleric and bibliophile). Bequeathed to the library of the Barefoot Carmelites of the convent of Saint Joseph of Barcelona. Entered in the library of the university after the suppression of the religious orders in 1835.

Madrid, Biblioteca Nacional de España

Shelf mark: 3/49075. Binding: Leather. Provenance: Manuscript annotation on title page: "Os justi meditabitur sapientiam." "Es del Doctor Fernando Ynfante de Aurioles [Fernando Infante de Aurioles] (seventeenth century | Spanish medical doctor) costole 88 Res. Año 1653." Library stamp: BR [Biblioteca Real].

Madrid, Museo Naval

Shelf mark: M-MN, CF-343. Binding: Mottled calf (Pasta Española).

Madrid, Real Academia de la Historia

Shelf mark: M-RAH, 5/2111. Binding: Parchment.

Madrid, Universidad Complutense de Madrid, Biblioteca

Copy 1: Shelf mark: BH MED 3257. Binding: Contemporary parchment over paper boards. Coloration: Black-and-white. Complete: Yes. Provenance: Library stamp: Real Colegio de Cirurgía de San Carlos de Madrid. Manuscript annotation on endpaper: Du Moulinis. Manuscript annotation: faint text on back cover. Manuscript annotation: crossed out name on title page and folium *2.

Copy 2: Shelf mark: BH FG 2751. Binding: Contemporary parchment over paper boards. Coloration: Black-and-white. Complete: Yes. Provenance: From the library of Francisco Guerra (1916–2011 | Spanish medical historian). Bequeathed to Complutense University in 2006.

Salamanca, Universidad de Salamanca

Copy 1: Shelf mark: BG/47074. Binding: Mottled calf (Pasta Española), spine with golden decorations and a red label with title in gold, edges in red. Coloration: Black-and-white. Complete: Yes.

Copy 2: Shelf mark: BG/41020. Binding: Contemporary parchment. Coloration: Black-and-white. Complete: Yes. Provenance: Manuscript annotation (Spanish) on title page: From the library of the Colegio Mayor de Cuenca. Acquired by the library early nineteenth century.

Valencia, Universidad de Valencia. Biblioteca Histórica

Shelf mark: BH Y-28/004. Binding: Parchment.

Sweden

Lund, Universitetsbiblioteket

Shelf mark: Fol Utl Naturv geogr [Brasilien]. Binding: Contemporary parchment. Coloration: Black-and-white. Complete: Yes. Provenance: Manuscript annotation on folium *2: [Illegible].

Norrköping, Stadsbibliotek

Collection: Finspongssamlingen Shelf mark: 1464 Fol. Binding: Contemporary parchment, five raised bands. Coloration: Black-and-white. Complete: Yes. Provenance: Library Stamps on flyleaf and title page: Finspongs Bibliotek and Finspongs Bibliothec. From the collection of the family De Geer (Family of industrialists from Walloon origin, belonging to Dutch and Swedish nobility. Probably acquired at an auction in the eighteenth or nineteenth century.

Stockholm, Karolinska Institutet, Hagströmerbiblioteket

Copy 1: Shelf mark: Hylla 38, 61. Binding: Contemporary calf, six raised bands, double gilt fillet border around sides. Coloration: Black-and-white. Complete: Yes. Provenance: Ex libris: C.D. Carlsson. Donated by C.D. Carlsson to Apotekarsocietetens Bibliotek.

Copy 2: Shelf mark: Fol. -1800 54 KIB. Binding: Contemporary vellum, fillets around the boards, stamped cartouches on upper and lower boards, stamped spine decorations. Coloration: Black-and-white. Complete: Yes. Provenance: Manuscript annotation by Anders Johan Hagström[er] (1753–1830) on title page: "Tillhör Carolinska Medico Chirurgiska Institutet. 1a 17 Sk: 1 hyll." Part of the collection that was donated by Abraham Bäck (1713–1795 | Swedish archiater) to Collegium Medicum where Bäck was president. Note: Annotated in two old hands. This is probably the copy that belonged to Abraham Bäck and was borrowed and possibly annotated by Daniel Rolander (1725–1793 | Swedish naturalist) when he went on his trip to Suriname, mentioned in a letter from Rolander to Bäck on 20 May 1756.

Copy 3: Shelf mark: Fol Rundet Eugenia. Binding: Contemporary alum-tawed pigskin over laminated boards, five raised double bands. Blind-tooled frames with center and corner pieces. Red edges. Traces of two paired ties of green textile at fore-edge. A leaf tab marker indicates the beginning of Marcgraf's work (parchment strip pasted at fore-edge). Coloration: Black-and-white. Complete: Yes. Provenance: Bookplate on pastedown: Werner Olsson. Stamp on pastedown: Föreningen Medicinhistoriska Museets vänner. Owner's signature on flyleaf: H.A. Eurén.

Stockholm, Kungliga Biblioteket

Copy 1: Shelf mark: RAR: 148 A Fol. Historia. Binding: Contemporary parchment, gilded coat of arms on front and back cover. Coloration: Black-and-white. Complete: Yes.

Copy 2: Shelf mark: Elz. 524 Fol. Binding: Nineteenth-century half parchment, marbled boards with monogram GJB. Binding identical to Elz. 525 Fol. Coloration: Black-and-white. Complete: Yes.

Copy 3: Shelf mark: Elz. 525 Fol. Binding: Nineteenth-century half parchment, marbled boards with monogram GJB. Binding identical to Elz. 524 Fol. Coloration: Black-and-white. Complete: Yes.

Stockholm, Universitetsbiblioteket

Collection: Bergianska Biblioteket. Shelf mark: H.V:2.2.n.1. Binding: Contemporary calf, gold tooling. Coloration: Black-and-white. Complete: Yes. Provenance: Wax seal: Magnus Gabriel De la Gardie (1622–1686 | Swedish statesman and military commander). From the collection of Bengt Bergius (1723–1784 | Swedish book collector) and Peter Jonas Bergius (1730–1790 | Swedish medical doctor). Bequeathed to the Royal Swedish Academy of Sciences in 1790. Deposited at Stockholm University Library.

Uppsala, Universitetsbiblioteket

Copy 1: Shelf mark: Nat.vet. Allm. Fol. [Piso] (Ex.: 1). Binding: Half-parchment with paper boards (defective), paper title-label with printed text on spine, which reads G.Pisonis | et. G.Marc. | Gravi. Histor | Natur. Brasiliae. Coloration: Black-and-white. Complete: no, lacks pages 281–293. Provenance: Ex libris: Johan Lindestolpe (1678–1724 | Swedish medical doctor and botanist).

Copy 2: Shelf mark: Nat.vet. Allm. Fol. [Piso] (Ex.: 2). Binding: Contemporary parchment. Coloration: Black-and-white. Complete: Yes.

Copy 3: Collection: Leufsta collection. Shelf mark: Leufstasaml. F 92. Binding: Contemporary parchment. Coloration: Black-and-white. Complete: Yes. Provenance: Manuscript annotation on flyleaf, which reads: Tutciari numina Noriana Villa | Gratissima omnium flora | ut purpureum viridi gentium de cespite florem | perpetuo efflorescat, latiquer fontis peren- | nitate foecundet | cum voto | Confaer. | G.P.I. From the collection of Baron Charles De Geer (1720–1778 | Swedish industrialist and entomologist). Bookplate: Uppsala Universitets Bibliotek, Leufsta samlingen.

Switzerland

Basel, Universitätsbibliothek

Shelf mark: Hx I 1. Binding: Contemporary leather, traces of ribbons. Coloration: Black-and-white. Complete: Yes. Provenance: Manuscript annotation: [illegible name] 1650. Manuscript annotation on half-title: J.J. d'Annone. 1753. From the collection of Johann Jacob d'Annone (1728–1804 | Swiss jurist, archeologist, collector of books and naturalia).

Bern, Universitätsbibliothek

Location: Bibliothek Münstergasse. Shelf mark: MUE Gross W 49. Binding: Contemporary parchment. Coloration: Black-and-white. Complete: Yes. Provenance: Bookseller's ticket on front pastedown: Huber & Comp. (Hans Körber) Buch- u. Kunsthandlung in Bern, Kramgasse 141. Probably acquired by the Municipal Library of Bern between 1864 and 1892.

Zürich, Zentralbibliothek

Copy 1: Shelf mark: NNN 64 | F. Binding: Contemporary parchment with gilded supralibros of Der Naturforschenden Gesellschaft in

Zürich. Coloration: Black-and-white. Complete: Yes. Provenance: From the collection of Der Naturforschenden Gesellschaft in Zürich, founded in 1746, and probably its predecessor the Physikalische Gesellschaft.

Copy 2: Shelf mark: TZ 61 | G.

United Kingdom

Aberdeen, University Library

Copy 1: Shelf mark: SB f590 Pis.

Copy 2: Shelf mark: f KCx 59 Pis.

Copy 3: Shelf mark: FL f Zeta 2.49.

Blickling, Blickling Hall National Trust Library

Shelf mark: 5513. Binding: Eighteenth-century sprinkled calf; gilt double fillet border; gold-tooled spine, with gilt title. Red and brown sprinkled edges. Coloration: Black-and-white. Complete: Yes. Provenance: From the collection of: Sir Richard Ellys (1682–1742 | English politician and book collector). Manuscript initial on front flyleaf: "M." [i.e. catalogue code of John Mitchell (ca. 1685–1751 | librarian to Sir Richard Ellys). Manuscript annotation: Eighteenth-century annotations relating to purchase on flyleaf. Acquired in 1940 with the whole of the Blickling Hall estate by gift in the bequest of Philip Kerr, 11th Marquis of Lothian (1882–1940 | British politician, diplomat, and newspaper editor).

Cambridge, Balfour & Newton Library

Shelf mark: 5 folio 80. Provenance: Bookplate: Bequeathed by Alfred Newton (1829–1907 | English zoologist and ornithologist) to the Museum of Zoology.

Cambridge, Gonville & Caius: Lower Library

Shelf mark: L.19.11.

Cambridge, St. Catharine's College Library

Shelf mark: Addenbrooke.L.2.1. Provenance: Donated to the library in 1718 by John Addenbrooke (1680–1719 | English medical doctor).

Cambridge, St. John's College: Upper Library

Shelf mark: Mm.1.24.

Cambridge, University Library

Copy 1: Collection: Part of the "Stars" (historic collection). Shelf mark: N*.1.30(B). Binding: Seventeenth-century sprinkled calf, blind ruled lines to the boards. Coloration: Black-and-white. Complete: Yes. Provenance: Bequeathed by Henry Lucas (c. 1610–1663 | English clergyman and politician) in 1664.

Copy 2: Collection: The "Royal" Library. Shelf mark: L.1.9 (OS). Binding: Seventeenth-century leather, gold tooling to the boards. Rebacked. Coloration: Black-and-white. Complete: Yes. Provenance: From the collection of John Moore, Bishop of Ely (1646–1714 | English clergyman and bibliophile), purchased and given to Cambridge University Library by King George I in 1715.

Durham, University Library

Collection: Ushaw College. Shelf mark: Ushaw XVIII.E.5.15. Binding: Eighteenth-century calf over boards, gold fillets; front board missing. Coloration: Black-and-white. Complete: No, lacks engraved title page. Provenance: Illegible circular purple ink stamp on letterpress title page.

Edinburgh, National Library of Scotland

Shelf mark: Am.1.17. Complete: no, second part only.

Edinburgh, University of Edinburgh Library

Shelf mark: JY493.

Edinburgh, Royal Botanic Garden Edinburgh

Shelf mark: F P.4. Binding: Contemporary vellum. Coloration: Black-and-white. Complete: Yes. Provenance: Bookplate on front pastedown: John Hope M.D. (1725–1786 | Scottish physician and botanist). Bookplate on verso of title page: Jo. [John] Stevenson, M.D. (Scottish medical doctor, father-in-law of John Hope).

Edinburgh, Royal College of Physicians of Edinburgh

Collection: Strong room. Shelf mark: SS 2.21. Binding: Blind-tooled boards.

Leeds, University Library

Location: Brotherton Library Special Collections. Shelf mark: Strong Room for. fol. 1648 PIS.

London, British Library

Copy 1: Shelf mark: 443.k.7. Binding: Rebound in green half leather with green cloth boards. Coloration: Colored. Complete: Yes. Provenance: Manuscript annotation on title page: Bibliotheca Sloanianæ Min: 114. From the collection of Sir Hans Sloane (1660–1753 | Anglo-Irish physician, naturalist, and collector). Note: No annotations, probably a presentation copy.

Copy 2: Shelf mark: 443.k.8. Binding: Rebound in half calf with brown cloth boards. Coloration: Black-and-white. Complete: Yes. Provenance: Manuscript annotation on title page: Ex Libris Jacobi Petiver Pharm. London [James Petiver] (c. 1665–1718 | English apothecary). Acquired by Sir Hans Sloane in 1718. Note: Annotated throughout, contains inserted letters addressed to or written by James Petiver.

Copy 3: Shelf mark: 452.g.8. Binding: Rebound in half calf with brown cloth boards, spine with six raised bands, blind stamped on front cover. Coloration: Black-and-white. Complete: Yes. Provenance: Stamp: Joseph Banks (1743–1820 | English naturalist and botanist).

Copy 4: Shelf mark: 37.g.18. Binding: Half-leather with marbled boards and marbled edges, supralibros of King George III that was used between 1801 and 1816. Coloration: Black-and-white. Complete: Yes. Provenance: From the Royal collection of George III (1738–1820 | King of Great Britain and Ireland).

London, Kew Gardens Library

Shelf mark: PRE-LINN-C PIS. Binding: Leather. Coloration: Black-and-white. Complete: Yes. Provenance: Bookplate: Royal Gardens Kew, presented by George Bentham (1800–1884 | English botanist) 1854.

London, The Linnean Society of London

Shelf mark: T13661. Binding: Contemporary parchment with later printed green title-labels. Coloration: Black-and-white. Complete: Yes. Provenance: Manuscript annotation on title page: Ex bibl. Linn. 1784 J.E. Smith. The library of Carl Linnaeus (1707–1778 | Swedish botanist and taxonomist) was bought as a whole by James Edward Smith

(1759–1828 | English botanist and founder of the Linnean Society) from the widow of Linnaeus in 1784. Note: Generic identification to some botanical figures and a manuscript annotation on front flyleaf by Linnaeus.

London, Middle Temple Library

Shelf mark: BAY L529. Binding: Probably eighteenth-century leather, gold tooling. Coloration: Black-and-white. Complete: Yes, title page defective.

London, Natural History Museum

Copy 1: Location: South Kensington. Collection: Botany Special Collections. Shelf mark: Special Books 581.9(81) PIS F. Binding: Modern black buckram binding with black leather spine and corners. Coloration: title page: colored; other illustrations: Black-and-white. Complete: no, 2nd section by Marcgraf lacks pages 3–6. Provenance: Armorial bookplate: [not identified].

Copy 2: Location: South Kensington. Collection: General Special Collections. Shelf mark: 4 f PIS. Binding: Contemporary vellum. Coloration: Black-and-white. Complete: Yes. Provenance: Manuscript annotation: [illegible name].

Copy 3: Location: Tring, Hertfordshire. Collection: Rothschild Collection. Shelf mark: ROTHSCHILD LIB. 81/F. Binding: Contemporary vellum with tooling and raised bands on the spine. Coloration: Black-and-white. Complete: no, lacks title page. Provenance: Lionel Walter Rothschild, 2nd Baron Rothschild, Baron de Rothschild (1868–1937 | British zoologist).

London, Royal College of Physicians

Copy 1: Location: Dorchester Library. Shelf mark: D1/28-e-3. Binding: Leather with a simple panel design and gold tooling, rebacked, red-speckled edges. Coloration: Black-and-white. Complete: Yes. Provenance: Bookplate: Royal College of Physicians. Possibly from the library of Henry Pierrepont, Marquis of Dorchester (1606–1680 | English peer), bequeathed as a whole to the college in 1680. Note: Fragment of original front flyleaf with old shelf number "E4;" pencil annotation Christopher Merret (1614/5–1695 | English physician and scientist, first librarian of the Royal College of Physicians) suggests that this copy was in the library before the Great fire of 1666, yet this contradicts with the attribution of this copy to the legacy of the Marquis of Dorchester.

Copy 2: Location: Dorchester Library. Shelf mark: D1/9-f-2. Binding: Contemporary blind-stamped parchment with double panel design. Coloration: Colored. Complete: Yes. Provenance: Bookplate: Royal College of Physicians. From the library of Henry Pierrepont, Marquis of Dorchester, bequeathed as a whole to the college in 1680.

London, Royal Society Library

Shelf mark: Medicine-large. Binding: Twentieth-century half-leather, buckram. Coloration: Black-and-white. Complete: Yes.

London, Wellcome Library

Shelf mark: 41391/D. Binding: Twentieth-century leather. Coloration: Black-and-white. Complete: Yes. Provenance: Manuscript annotation on title page: [illegible]. Acquired by the Wellcome Library around 1947.

London, Zoological Society of London Library

Shelf mark: 2ICP. Binding: Nineteenth-century half-leather with marbled boards, brown-speckled edges. Coloration: Black-and-white. Complete: Yes.

Manchester, University Library

Location: John Rylands Library. Shelf mark: 1943.

Oxford, Bodleian Library

Copy 1: Shelf mark: Locke 16.5. Binding: Calf with gold rules. Complete: Yes. Provenance: Manuscript annotation: John Locke (1632–1704 | English philosopher).

Copy 2: Shelf mark: Lister D 68. Binding: Calf. Complete: Yes. Note: Folium L2 missigned L3.

Copy 3: Shelf mark: Vet. B3 b.26. Binding: Half calf with marbled paper boards. Rebacked. Complete: no, lacks folia *2-3. Provenance: Bookplate: Radcliffe Library. Note: Large-paper copy.

Copy 4: Shelf mark: R 1.16 Med. Binding: Mottled calf with blind rules. Complete: Yes.

Copy 5: Shelf mark: M 1.6 Med. Binding: Sprinkled calf with blind rules. Complete: Yes.

Oxford, Jesus College Library

Collection: Fellows' Library. Shelf mark: L.7.10 Gall. Complete: Yes. Provenance: Bookplate: "Hunc librum olim suum Collegio Iesu legauit Eduardus Herbert Baro De Cherbury A.D. MDCXLVIII" on inside of upper board [Edward Herbert of Cherbury] (1583–1648 | Anglo-Welsh soldier, diplomat, historian, poet, and religious philosopher).

Oxford, Magdalen College Library

Collection: Old Library. Shelf mark: R.18.4. Binding: Tight-back binding: tanned medium-brown calf over pasteboard. Decoration: blind-tooled, triple-fillet outer frame. Spine: five raised bands: (panel 2) gold-tooled leather: MAURITII HIST. NAT. BRASILIAE; (panels 1 and 6) blind-tooled hatching/fillets; (panel 5) gold- tooled 1648; (panel 6) gold-tooled Goodyer canting crest and gold-tooled no. 53. Coloration: Black-and-white. Complete: Yes. Provenance: Donated to the library by John Goodyer (c.1592–1664 | English botanist and physician). Manuscript annotation: price in the hand of John Goodyer: 1l.-2s.

Oxford, Museum of Natural History Library

Collection: Hope Library (Entomology). Shelf mark: Folios: E.37.

Oxford, Sherardian Library

Collection: Herbarium store. Shelf mark: Sherard 665/BT.

Oxford, St John's College Library

Shelf mark: Y.1.2. Binding: Seventeenth or eighteenth-century stiffened parchment. Complete: Yes. Provenance: Bookplate: STJ type 4 (foliate facsimile). Manuscript annotation on flyleaf: J. [John] Merrick (1670–1757 | English medical doctor).

Windsor, Eton College

Shelf mark: Ac.2.13. Binding: Contemporary gilt leather. Coloration: Black-and-white. Complete: Yes. Provenance: Label recording bequest to Eton by Nicholas Mann (d. 1753 | British antiquary and Master of the Charterhouse) in 1754 on front pastedown. Previous shelf marks: Dd.1.6 and Aa.2.09.

Wroughton, Science Museum Library

Shelf mark: F O.B. PIS PISO.

United States

Baton Rouge, LA, Louisiana State University Library

Collection: McIlhenny Natural History Collection. Shelf mark: McIlhenny Flat QH117 .P67. Binding: Brown leather over boards. Coloration: Black-and-white. Complete: Yes. Provenance: Paper label on flyleaf wit manuscript annotation: 10 dec. 1682, 14. H. Lud. Morin. D.M.P. [Louis Morin de Saint-Victor] (1635–1715 | French medical doctor and botanist). From the collection of Edward Avery McIlhenny (1872–1949 | American businessman and conservationist). Purchased by the library from a rare book dealer in 2010.

Bethlehem, PA, Lehigh University Libraries

Location: Linderman Library. Shelf mark: 574.981 P678h.

Bloomington, IN, Indiana University

Location: Lilly Library. Collection: Mendel collection. Shelf mark: QH117 .P67. Binding: Mottled calf with speckled edges. Coloration: Black-and-white. Complete: Yes. Provenance: From the library of Charles Ralph Boxer (1904–2000 | British historian of Dutch and Portuguese colonial history).

Boston, MA, Boston Public Library

Shelf mark: RARE BKS XfL.648 .P67H pt.1-2. Binding: Contemporary parchment. Coloration: Black-and-white. Complete: Yes. Provenance: Manuscript annotation (Dutch): Given by Maria Bevelot of Terschelling (d. 1713? | Dutch governess of Terschelling) to Jacob Grevenstein (c.1670–1711 | Dutch medical doctor). Bookplate: Property of the Public Library of Boston, from the Bates fund, added Feb. 5, 1862.

Cambridge, MA, Harvard Library

Copy 1: Location: Gray Herbarium. Shelf mark: Botany Gray Herbarium Oversize Ka P67h 1648. Binding: Probably contemporary vellum, tooling on front and back cover. Coloration: Black-and-white. Complete: Yes. Provenance: Stamp: Adl. Dominium Metgethen. Accessioned 25 February 1911.

Copy 2: Location: Arnold Arboretum. Shelf mark: Botany Arnold (Cambr.) Oversize Ka P67h 1648. Binding: Vellum, rebound with the original cover affixed to the newer binding, tooling on front and back cover. Coloration: Black-and-white. Complete: Yes.

Copy 3: Location: Houghton Library. Shelf mark: Typ 632.48.695.

Copy 4: Location: Houghton Library. Shelf mark: F 5390.10.5.

Copy 5: Location: Dumbarton Oaks. Shelf mark: RBR K-3-2 PIS. Provenance: Bookplate: Waldemar Schwalbe, 1937.

Copy 6: Location: Ernst Mayr Library of the Museum of Comparative Zoology. Shelf mark: Spec. Coll.

Chapel Hill, NC, University of North Carolina

Shelf mark: QH117 .P67. Binding: Contemporary brown calf, gold and blind fillets around front and back, gold lines at bands (sewn on double cords), small center tool on each panel, title in gold on second panel. Coloration: Black-and-white. Complete: Yes. Provenance: Bookplate, front flyleaf and verso contained within decorative paper borders applied by unknown persons. Supplemental materials included in pocket on rear pastedown. Page of manuscript tipped in preceding title page.

Chicago, IL, Field Museum Library

Collection: Ayer collection. Shelf mark: Ayer add.2 1648.1*. Binding: Eighteenth-century mottled calf, blind tooling to the boards, spine restored. Coloration: Black-and-white. Complete: Yes. Provenance: Part of the donation by Edward E. Ayer (1841–1927 | American business magnate and philanthropist) to the Field Museum in 1926.

Chicago, IL, Newberry Library

Shelf mark: Ayer 1269 .B8 P67 1648. Binding: Eighteenth century half-leather. Gilt letterpress text on burgundy and black leather title-labels. Coloration: Black-and-white. Complete: No, lacks pp. 102–104. Provenance: Engraved title-page: Paris Novemb: 25 1669. Manuscript name at the end of the text: Pa. Moray. Bookplate: presented to the library by Edward Everett Ayer, 1911. Purchased by the library for the Edward A. Ayer collection in May 1951. Note: botanical sections annotated in old hand throughout, probably second half eighteenth century (binominal nomenclature). Half-title Georgi Marcgravi de

Liebstad ... Historiæ rerum naturalium Brasiliæ, libri octo with two pencil portrait sketches at the bottom of the page.

Chicago, IL, University of Chicago

Location: John Crerar Collection of Rare Books in the History of Science and Medicine. Shelf mark: f QH117.P670 c.1. Binding: Twentieth-century library binding. Coloration: Black-and-white. Complete: Yes. Provenance: Bought by the library in the twentieth century.

Cincinnati, OH, Lloyd Library and Museum

Shelf mark: QH117 .P67 1648 RBR (ff). Binding: Paper-covered boards. Coloration: Black-and-white. Complete: no, lacks title page.

Columbia, MO, University of Missouri

Shelf mark: MU Ellis Special Collections Rare Vault QH117 .P67. Binding: Vellum. Coloration: Black-and-white. Complete: Yes.

Hartford, CN, Trinity College, Watkinson Library

Shelf mark: Special Quarto QH117 .P67. Binding: Rebound in the twentieth century, quarter bound in red morocco. Coloration: Black-and-white. Complete: Yes. Provenance: Bought by the library from Henry Stevens (1819–1886 | American bibliographer and book agent), London, in 1868.

Ithaca, NY, Cornell University Library

Shelf mark: QH117 .P67.

Kansas City, MO, Linda Hall Library of Science, Engineering and Technology

Shelf mark: QH117 .P5 1648 folio. Binding: Contemporary vellum. Coloration: Black-and-white. Complete: Yes. Provenance: Library stamp: Manuscript annotation: Bornius. Ex libris Edinensis Medicae Societatis [Royal Medical Society, Edinburgh]. Bought by the library from a bookdealer in 1983.

Lawrence, KS, University of Kansas, Kenneth
Spencer Research Library

Collection: Solon E. Summerfield Collection. Shelf mark: Summerfield G196. Coloration: Black-and-white. Complete: Yes. Provenance: Bought by the library from bookseller HP Kraus in March 1968.

Los Angeles, CA, Getty Research Institute

Shelf mark: QH117 .P67. Binding: Eighteenth-century brown calf, marbled endpapers. Coloration: Black-and-white. Complete: Yes. Provenance: Bookplate: C.J.L. [Charles Jacques Louis] Coquereau (1744–1796 | French medical doctor). Bookplate: Theodore [Deodatus Nathaniel] Besterman (1904–1976 | researcher, bibliographer, and collector).

Los Angeles, CA, University of California Library

Copy 1: Shelf mark: BIOMED *QH 117 P676h 1648 RARE. Binding: Vellum over boards, with yap edges. Coloration: Black-and-white. Complete: Yes. Provenance: Manuscript annotation on title page: [illegible]. Manuscript annotation on upper free endpaper: "Tho White" and "Fol. 102." His monogram bookplate on upper paste-down.

Copy 2: Collection: Clark Library. Shelf mark: f QH117 .P67 *. Binding: Sprinkled calf, rebacked. Provenance: Armorial bookplate: Ormathmaite[?].

Madison, WI, University of Wisconsin

Shelf mark: 762187 noncurrent oversize. Binding: Contemporary vellum. Coloration: Black-and-white. Complete: Yes. Provenance: Purchased by the library in 1951.

Minneapolis, MN, University of Minnesota

Collection: James Ford Bell Library. Shelf mark: 1648 fPi. Binding: Contemporary limp vellum. Coloration: Black-and-white. Complete: Yes. Provenance: Purchased by the library from a rare book dealer in 1976.

New Haven. CT, Yale University Library

Shelf mark: S61h O58.

New Orleans, LA, Tulane University

Collection: Howard Tilton Memorial Library. Shelf mark: Latin American Library (Rare Oversize) QH117 .P67.

New York, NY, American Museum of Natural History

Shelf mark: C-6. Binding: Rebound in twentieth-century red buckram. Originally bound in mottled boards. Coloration: Black-and-white. Complete: Yes. Provenance: Accessioned in 1932, acquisition before that time.

New York, NY, Columbia University Libraries

Copy 1: Shelf mark: BookArt Z232.EL9 1648 P67 (copy one). Binding: Eighteenth-century brown leather, marbled endpapers, gold tooling to the spine. Coloration: Black-and-white. Complete: Yes. Provenance: Purchased with the library of the American Type Founders Company, 1941.

Copy 2: Shelf mark: BookArt Z232.EL9 1648 P67 (copy two). Binding: Modern library binding, rebound in 1949. Coloration: Black-and-white. Complete: no, lacks several folia. Provenance: Manuscript annotation: [illegible]. Purchased from Bangs, 10 March 1891.

New York, NY, New York Botanical Garden

Location: LuEsther T. Mertz Library. Shelf mark: fQH117 P5. Binding: Rebound in buckram in 1942, edges gilded. Coloration: Black-and-white. Complete: Yes. Provenance: Purchased by the library in 1902.

New York, NY, New York Academy of Medicine

Shelf mark: Folio Vault. Binding: Contemporary limp vellum with yapped edges. Coloration: Black-and-white. Complete: Yes. Provenance: Manuscript annotation on title page: Conventus Andreoviensis.

New York, NY, Public Library

Copy 1: Shelf mark: *KB+ 1648 (Piso, W. Historia natvralis Brasiliae) Copy 1. Binding: Contemporary vellum over pasteboard. Coloration: Black-and-white. Complete: Yes. Provenance: Bookplate: I.M.W. Baumann. Bookplate: C.E. [Christian Erhard] Kapp (1739–1824 | German medical doctor and translator).

Copy 2: Shelf mark: *KB+ 1648 (Piso, W. Historia natvralis Brasiliae) Copy 2. Binding: Later, probably nineteenth-century, calf over pasteboard with scoring and damage. Rebacked. Coloration: Colored. Complete: Yes.

Notre Dame, IN, Hesburgh Library

Copy 1: Shelf mark: Special Coll. Rare Books XLarge QH 117 .P676 1648. Binding: Marbled paper boards. Coloration: Black-and-white. Complete: Yes.

Copy 2: Shelf mark: Special Coll. Rare Books XLarge QH 117 .P676 1648. Binding: Modern binding. Original brown leather board with gold tooling enclosed in box. Coloration: Black-and-white. Complete: Yes. Provenance: Manuscript annotations in margins by Edward Lee Greene (1843–1915 | American botanist).

Philadelphia, PA, Academy of Natural Sciences of Drexel University Library & Archives

Shelf mark: Wolf Room Folio QH117 .P67. Binding: Vellum. Coloration: Black-and-white. Complete: Yes. Provenance: Donated to the library by Academy member John Howard Redfield (1815–1895 | American botanist) between 1861 and 1881.

Philadelphia, PA, University of Pennsylvania Libraries

Copy 1: Location: Kislak Center for Special Collections, Rare Books and Manuscripts Collection: Elzevier Collection Shelf mark: Elz F 3631. Binding: Full leather, boards detached, gold tooling on spine and gold double lines made with a fillet on edges of boards. Provenance: Donated to the University as part of the Elzevier Collection by E.B. [Edward Bell] Krumbhaar (1882–1966 | American pathologist).

Copy 2: Location: Pennsylvania Hospital Library.

Pittsburgh, PA, Carnegie Mellon University, Hunt Institute for Botanical Documentation

Shelf mark: +BD4 P678m. Binding: Seventeenth- or eighteenth-century mottled brown morocco with metal clasps, metal corners on the bottom, edges mottled green, red, and brown. Coloration: Black-and-white. Complete: Yes. Provenance: Manuscript annotation: Collegij Societatis Jesu Monasterij Westph. 1677. Library stamp: Ex. Bibl.

Paulina Monast. Library stamp: Ausgeschieden aus der Universitäts-Bibliothek Münster. Acquired by the Jesuit College in Münster in 1677. Transferred to the library of the Westfälische Wilhelms-Universität in 1780 after the dissolution of the Jesuit order in 1773. Later removed from the collection of that library. From the private collection of Rachel McMasters Miller Hunt (1882–1963 | American bookbinder and collector), donated as a whole to Carnegie Institute of Technology, now Carnegie Mellon University. Reference: This copy is no. 244 in the *Catalogue of Botanical Books in the Collection of Rachel McMasters Miller Hunt* (Pittsburgh, PA: Hunt Botanical Library, 1958). Note: The man and woman on the engraved title page are covered with a leaf.

Providence, RI, Brown University

Copy 1: Location: John Carter Brown Library. Shelf mark: 2-SIZE F648 .P678h. Binding: Contemporary vellum. Coloration: Black-and-white. Complete: Yes. Provenance: Bookplate: Prince Augustus Frederick, Duke of Sussex (1773–1843). Library stamp on folium *2: John Carter Brown.

Copy 2: Location: John Hay Library, Lownes room. Shelf mark: 2-SIZE QH117.P67.

San Francisco, CA, California State Library

Collection: Sutro Library. Shelf mark: 508.8 P67h. Note: Copy is missing.

St. Louis, MO, Missouri Botanical Garden Library

Copy 1: Collection: Sturtevant Pre-Linnean collection. Shelf mark: QH117 .P57 1648 c.1. Binding: Twentieth-century half-leather with red stained text blocks edges. Coloration: Colored. Complete: no, lacks folia †4 and 2K3.

Copy 2: Collection: Sturtevant Pre-Linnean collection. Shelf mark: QH117 .P57 1648 c.2. Binding: Contemporary pigskin. Coloration: Black-and-white. Complete: Yes. Provenance: Donated to the library in 1892 by Edward Lewis Sturtevant (1842–1898 | American agronomist and botanist). Manuscript annotations by Sturtevant. Manuscript annotation: Johann Heinrich[?]. Manuscript annotation: [illegible].

Urbana, IL, University of Illinois at Urbana Champaign

Shelf mark: Q570.981 P676h. Binding: Leather, tooling to the boards. Coloration: Black-and-white. Complete: Yes.

Washington, DC, The Catholic University of America

Collection: Oliveira Lima Library. Shelf mark: RBK 1129 1648. Binding: Contemporary vellum over boards. Coloration: Black-and-white. Complete: Yes. Provenance: Bookplate on front paste-down: M. de Oliveira Lima. Donated to the library by Manoel de Oliveira Lima (1867–1920 | Brazilian diplomat, journalist, and historian) in 1920.

Washington, DC, Folger Shakespeare Library

Shelf mark: 245- 321f. Binding: Half vellum over marbled paper boards, gold-stamped label and gold tooling on spine, marbled endpapers; edges stained red. Modern manuscript leaf with bibliographical note affixed to front free endpaper. Provenance: Bookplate: Ex libris Jacobi P.R. [James Patrick Ronaldson] Lyell (1871–1948 | British book collector). Bookplate: Mary P. Massey. Donated by Massey as part of collection of over 300 herbals to the Folger Shakespeare Library in 1994.

Washington, DC, Library of Congress

Copy 1: Shelf mark: QH117 .P67 Jefferson Exhibit Coll fol. Binding: Stamped in gilt on cover: Ober Rath. Coloration: Black-and-white. Note: The book is missing.

Copy 2: Shelf mark: QH117 .P67 fol. copy 2. Binding: Later buckram. Coloration: Colored (some of the pages with colored images were accidentally glued to each other because of the paint; in the process of opening those pages, the images were damaged). Complete: Yes. Provenance: Gift from James Carson Brevoort (1818–1887 | American book collector) of Brooklyn, NY, to the Smithsonian Library between 3 and 5 October 1885. The Smithsonian Library gave it to the Library of Congress, where it was accessioned on 17 October 1885.

Copy 3: Shelf mark: Rosenwald 1433 Rosenwald Coll. Binding: Contemporary vellum, arabesque medallion. Coloration: Black-and-white. Complete: Yes. Provenance: Bookplate: Richard Joseph Sullivan Esquire (1752–1806 | British MP and writer). Bookplate: LJR [Lessing Julius Rosenwald] (1891–1979 | American businessman, book collector, and philanthropist). Bequeathed by Rosenwald to the Library of Congress in 1979.

Copy 4: Shelf mark: QH117 .P67 Pre-1801 Coll fol. Coloration: Black-and-white. Complete: Yes. Provenance: Bookplate.

Copy 5: Shelf mark: QH117 .P67 Kislak Coll fol. Binding: Contemporary vellum, arabesque medallion. Decorated paper pastedowns and endpapers. Coloration: Black-and-white. Provenance: Purchased at a Christie's auction on 17 March 1999. Donated to the library by the Jay I. Kislak Foundation. Note: Accompanying material laid in.

Washington, DC, Smithsonian Libraries

Location: Joseph F. Cullman 3rd Library of Natural History. Shelf mark: QH117 .P67X folio SCNHRB. Binding: Nineteenth-century half morocco (black) with single blind fillet, and marbled paper over boards; spine plain with five raised cords, lettering in gilt. Coloration: Black-and-white. Provenance: Bookplate: Jonathan Dwight (1858–1929 | American ornithologist). Bookplate: Smithsonian Institution Libraries. Gift of Marcia Brady Tucker (1884–1992 | American ornithologist). Mrs. Tucker acquired Dwight's ornithological library in the 1930s and donated it and her own bird books to SIL in 1970.

Vatican

Vatican City, Biblioteca Apostolica Vaticana

Copy 1: Shelf mark: Stamp.Barb.EEE.VII.26.

Copy 2: Shelf mark: R.G.Scienze.S.38.

Antiquarian Booksellers

Six copies were offered by antiquarian booksellers in 2018–2019 when this copy census was initially set up. Three years later, in June 2022, the listed copies of Antiquariaat Junk and Richard C. Ramer Old and Rare Books are no longer available and have probably been sold. A previously unlisted copy is now offered by Antiquariat Winfried Kuhn. At least eight copies, all of them uncolored, were auctioned in the past 25 years, we mention the copies sold at Christies (1999, 2003, 2004, 2009, 2014, and 2015) and at Bubb Kuyper (2019, 2020). The antiquarian copies have been added to the copy census, because their thorough descriptions generally allow future identification with copies that may appear elsewhere on the market or in institutions. The auctioned copies have been discarded from the copy census, since provenance data is often left out of the description and consequently the copy cannot always be identified indisputably.

Amsterdam, Antiquariaat Junk

Binding: Contemporary calf, gilt ornamented spine in seven compartments, sides with large gilt English Royal coat of arms of James II and the motto of the English chivalric Order of the Garter "Honi soit qui mal y pense," and two gilt borders. Coloration: Black-and-white. Complete: Yes.

Provenance: Supralibros of James II (1633–1701 | King of England and Ireland as James II and King of Scotland as James VII) as Duke of York, thus part of his collection before 1685. Bookplate: John Henry Gurney (1819–1890 | English banker, book collector, and ornithologist). Armorial bookplate: John Roland Abbey (1894–1969 | English book collector).

Berlin, Antiquariat im Hufelandhaus GmbH

Binding: Contemporary parchment. Coloration: Black-and-white. Complete: Yes. Provenance: North Library (dated 1860).

Berlin, Antiquariat Winfried Kuhn

Binding: Contemporary parchment. Coloration: Black-and-white. Complete: Yes.

New York, Arader Galleries

Binding: Contemporary calf, spine in seven compartments, with six raised bands. Coloration: Black-and-white. Complete: Yes. Provenance: Stamp: G[eorges] Cuvier (1769–1832 | French naturalist). With marginal annotations by Cuvier in the sections on fish and birds. Stamp: Muséum national d'Histoire naturelle, Paris. Auctioned at Christies, 19 November 2014.

New York, Richard C. Ramer Old and Rare Books

Binding: Contemporary morocco, eight-pointed star design in gilt with red straight-grained morocco inlay on front cover. Coloration: Black-and-white. Complete: Yes. Provenance: Manuscript annotation in upper blank margin of engraved title page: "Bibliothèque N.° 26." Manuscript annotation of a Benedictine monastery, dated 1664, on second leaf recto. Bookplate removed from front pastedown.

Paris, Librairie Camille Sourget

Binding: Contemporary blind-stamped pigskin over wooden boards. Coloration: Black-and-white. Complete: Yes.

West Newbury, MA, Barberry Hill Books

Binding: Half-parchment with blue paper boards. Coloration: Black-and-white. Complete: Yes.

Index

For Product Safety Concerns and Information please contact our EU
representative GPSR@taylorandfrancis.com
Taylor & Francis Verlag GmbH, Kaufingerstraße 24, 80331 München, Germany

www.ingramcontent.com/pod-product-compliance
Lightning Source LLC
Chambersburg PA
CBHW072133170526
45158CB00004BA/1360